Martin Breunig

On the Way to Component-Based 3D/4D Geoinformation Systems

With 106 Figures and 2 Tables

Springer

Author

Prof. Dr. Martin Breunig
Institute of Computer Science III
University of Bonn
53117 Bonn

Institute of Environmental Sciences
University of Vechta
49364 Vechta
Germany

E-mail: mbreunig@ispa.uni-vechta.de

"For all Lecture Notes in Earth Sciences published till now please see final pages of the book"

ISSN 0930-0317
ISBN 3-540-67806-9 Springer-Verlag Berlin Heidelberg New York

Cataloging-in-Publication data applied for
Die Deutsche Bibliothek - CIP-Einheitsaufnahme
Breunig, Martin:
On the way to component based 3D/4D geoinformation systems / Martin Breunig.
Berlin; Heidelberg; New York; Barcelona; Hongkong; London; Mailand; Paris; Singapur; Tokio: Springer 2001
(Lecture notes in earth sciences; 94)
  ISBN 3-540-67806-9

This work is subject to copyright. All rights are reserved, whether the whole or part of the material is concerned, specifically the rights of translation, reprinting, re-use of illustrations, recitation, broadcasting, reproduction on microfilms or in any other way, and storage in data banks. Duplication of this publication or parts thereof is permitted only under the provisions of the German Copyright Law of September 9, 1965, in its current version, and permission for use must always be obtained from Springer-Verlag. Violations are liable for prosecution under the German Copyright Law.

Springer-Verlag Berlin Heidelberg New York
a member of BertelsmannSpringer Science+Business Media GmbH

© Springer-Verlag Berlin Heidelberg 2001
Printed in Germany

The use of general descriptive names, registered names, trademarks, etc. in this publication does not imply, even in the absence of a specific statement, that such names are exempt from the relevant protective laws and regulations and therefore free for general use.

Typesetting: Camera ready by author
Printed on acid-free paper   SPIN: 10747604   32/3130   5 4 3 2 1 0

Lecture Notes in Earth Sciences   94

Editors:
S. Bhattacharji, Brooklyn
G. M. Friedman, Brooklyn and Troy
H. J. Neugebauer, Bonn
A. Seilacher, Tuebingen and Yale

Springer
*Berlin
Heidelberg
New York
Barcelona
Hong Kong
London
Milan
Paris
Singapore
Tokyo*

Preface

The ideas for this book were conceived during the last years when I was with GIS groups at the Technical University of Darmstadt (group of Hans-Jörg Schek) and at Bonn University (group of Armin B. Cremers). The results in the book are closely connected with two research projects that have been supported by the German Research Foundation (DFG): the collaborative research centre (SFB 350) "Interactions between and Modelling of Continental Geosystems" which was set up at Bonn University and the joint project "Interoperable Geoinformation Systems (IOGIS)".

First and foremost, I am very grateful to Armin B. Cremers for the supervision of my habilitation thesis and the support of the GIS group. In the same way I am indebted to Agemar Siehl for his support during many years of collaborative work and for reading the manuscript from a geological view point at short notice. Without him I would not have arrived at Bonn at all. I thank Rainer Manthey for his valuable hints and Horst Neugebauer for his engagement within the SFB 350. He motivated me to write the English manuscript.

I am also indebted to my colleagues Oleg Balovnev, Andreas Bergmann, Thomas Bode, Jörg Flören, Martin Hammel, Wolfgang Müller, Marcus Pant, Sergej Shumilov, Jörg Siebeck, Michael Werner from the Computer Science Institute and to Rainer Alms, Christian Klesper, Tobias Jentzsch, Robert Seidemann, Andreas Thomsen, Michael Klett, Thorsten Utescher from the Geological Institute as well as Hajo Götze and Sabine Schmidt from Free University of Berlin for many discussions and for the ever pleasant atmosphere. The trips to Berlin have completed my picture of a component-based GIS. Oleg did a great job during the GeoToolKit implementation.

I am grateful to my parents who encouraged me to write this book. Finally I am most obliged to thank Christine and Philipp for their continuous encouragement and patience during the completion of this book. To Philipp I am also indebted for lending me his rubbers, pencil-sharpeners and his modeling clay for the modeling of tetrahedra models. I thank Christine for her help and care in reading the manuscript. Last, but not least thank you to Springer for copy editing the english manuscript.

Bonn, June 2000   Martin Breunig

Contents

1 Introduction   1
- 1.1 Research Field "Geoinformation Systems" 1
- 1.2 Overview 2

2 Fundamental Principles   3
- 2.1 "Geoinformation Science" 3
  - 2.1.1 Change of the Notion 'GIS' 3
  - 2.1.2 GIS Research in Germany: a Retrospective 3
  - 2.1.3 International Development 4
  - 2.1.4 3D/4D Geoinformation Systems 6
  - 2.1.5 On the Way to Component-Based Geoinformation Systems 6
- 2.2 Geodata Modelling 8
  - 2.2.1 Peculiarities of Geoscientific Data: Spatial and Temporal Reference 8
    - 2.2.1.1 Pointset Topology 9
    - 2.2.1.2 Algebraic Topology 9
    - 2.2.1.3 Pointset and Algebraic Topology in Time 12
  - 2.2.2 Approaches of Spatial Data Models 13
    - 2.2.2.1 Field-Based Models 13
    - 2.2.2.2 Object-Based Models 15
  - 2.2.3 Temporal Extension of the Field-Based and the Object-Based Approach . 17
  - 2.2.4 Implementation of Spatial Data Models 17
    - 2.2.4.1 Vector Representation 17
    - 2.2.4.2 Raster Representation 17
  - 2.2.5 Layer Model 19
  - 2.2.6 GIS Types 20
    - 2.2.6.1 Vector-Based GIS 20
    - 2.2.6.2 Raster-Based GIS 21
    - 2.2.6.3 3D/4D Geoinformation Systems 21

	2.3	Geodata Management	22
		2.3.1 Functionality of a Geodatabase	22
		2.3.2 Spatial Access Methods	24
		2.3.2.1 Quadtree on Top of B^*-Tree	24
		2.3.2.2 Grid Files	28
		2.3.2.3 R-Trees	29
		2.3.3 From the Relational to the Object-Oriented Database Design..........	30
	2.4	Geodata Analysis	32
		2.4.1 Elementary Geometric Algorithms for GIS Applications	32
		2.4.2 Visualization of Spatial and Temporal Data	39

3 Examples of Today's Geoinformation Systems — 41

	3.1	Commercial Systems	41
		3.1.1 ARC/INFO	41
		3.1.2 SYSTEM 9	46
		3.1.3 SMALLWORLD GIS	47
	3.2	Research Prototypes	49
		3.2.1 GEO^{++}	49
		3.2.2 GeO$_2$	50
		3.2.3 GODOT	51
		3.2.4 GeoStore	52
	3.3	Comparison	54

4 Data Modelling and Management for 3D/4D Geoinformation Systems — 57

	4.1	Relevance of Space and Time	57
		4.1.1 Space	57
		4.1.2 Time	58
		4.1.3 Analogies between Space and Time	58
		4.1.3.1 Topological Properties	58
		4.1.3.2 Metric Properties	59
		4.1.3.3 Field-Based and Object-Based Modelling	60
		4.1.3.4 Approximation in Space and Time	60
		4.1.4 Differences	62

4.2	Modelling of Spatial and Temporal Objects		62
	4.2.1	Suitable Spatial and Temporal Representations	62
		4.2.1.1 Generalized Maps	63
		4.2.1.2 α-Shapes	65
		4.2.1.3 Convex Simplicial Complexes	66
	4.2.2	Spatial and Temporal Operations	67
		4.2.2.1 Classification	67
		4.2.2.2 Spatial Topological Relationships	69
		4.2.2.3 Temporal Change of the Spatial Topology	73
		4.2.2.4 Change of Geometry, Topology and Connected Components for Objects in Time	77
		4.2.2.5 Temporal Topological Relationships	83
	4.2.3	Geo-Objects in Space and Time	83
		4.2.3.1 Equality of Spatio-Temporal Objects	84
4.3	Management of Spatial and Temporal Objects		86
	4.3.1	Checking of Spatial and Temporal Integrity Constraints	86
		4.3.1.1 Checking of the Geometry	86
		4.3.1.2 Checking of the Topology	87
	4.3.2	Spatial and Temporal Database Queries	88
	4.3.3	Spatio-Temporal Database Access	92
	4.3.4	Supporting the Visualization of Large Sets of Spatio-Temporal Objects	98

5 Systems Development: from Geodatabase Kernel Systems to Component-Based 3D/4D Geoinformation Systems   **103**

5.1	Geodatabase Kernel Systems		103
	5.1.1	Requirements	103
	5.1.2	The DASDBS-Geokernel	104
		5.1.2.1 Concepts	104
		5.1.2.2 Architecture	105
		5.1.2.3 Coupling the DASDBS Geokernel with a Map Construction System	106
	5.1.3	The Object Management System	107
		5.1.3.1 Architecture	108
		5.1.3.2 Deep Embedding of 3D Data Types and Access Paths	108
		5.1.3.3 Database Query Language	109

5.2	The GeoToolKit		109
	5.2.1	Historical Development	110
	5.2.2	Component-Based Architecture	110
	5.2.3	Data Model	111
	5.2.4	Spatial Representations	114
	5.2.5	Spatial Indexes	115
	5.2.6	Extensible 3D Visualization	119
5.3	Example of an Application: Balanced Restoration of Structural Basin Evolution		121
5.4	Realization of Geological Components for a 3D/4D Geoinformation System		125
	5.4.1	Object Model Editor	125
	5.4.2	Integrity Checking Component	126
	5.4.3	Database Support of the Interactive Geological Modeling	131
	5.4.4	Support of Different Examination Areas	136
	5.4.5	Component for the Management of Time-Dependent Geologically Defined Geometries	137

6 Data and Methods Integration 143

6.1	Starting with the Geoscientific Phenomenon: the Meta Data Approach		143
	6.1.1	Retrieval of Meta Data	144
	6.1.2	Similarity Queries	144
6.2	Starting with the Data: the Original Data Approach		144
	6.2.1	Integrated Database Views: Spatial Operators for Schema Integration	145
	6.2.2	A Case Study	147
		6.2.2.1 Integration of Different Types of Well Data	147
		6.2.2.2 Integration of the Well Data with 3D Model Data	152
6.3	Integration of Heterogeneous Spatial Representations		153
	6.3.1	The Problem	153
	6.3.2	Spatial Representations Used in Geoscientific 3D Tools	154
6.4	An Example from Geology and Geophysics		155
	6.4.1	Objectives of the "Integrated Application"	155
	6.4.2	Consistent Object Exchange and Data Model Integration	157
	6.4.3	Geological-Geophysical 3D Object Model	159
	6.4.4	Methods Integration	161

7 Systems Integration   165

   7.1   Requirements 165

   7.2   Approaches for Systems Integration 166

   7.3   Coupling Mechanisms 168

           7.3.1   Client-Server Connection of an OODBMS 168

           7.3.2   Client-Server Connection via Standard UNIX Technology 169

           7.3.3   CORBA Communication Platform 169

           7.3.4   Persistency of CORBA Objects 170

   7.4   Implementation of a Component-Based Geoinformation System 172

           7.4.1   Access to an Open Geo-Database 172

           7.4.2   CORBA-Based Solution. 173

8 Outlook   179

References   181

Index   197

Chapter 1

Introduction

In this first chapter the latest developments of the research field "geoinformation systems" are discussed from the author's point of view. Furthermore, some explanations of notions used in the book, and an overview are given.

1.1 Research Field "Geoinformation Systems"

The research field "geoinformation systems" (GIS) has developed in an interdisciplinary fashion unlike most other fields. This obviously involves advantages, however, it also holds problems. Without any doubt one of the advantages is that a discourse between the various geoscientific disciplines, the engineering sciences and computer science has started. This encompasses the important questions of standardized representations for geo-objects, the management and the processing of geoinformation. Solutions to this problem improve the availability and the usability of spatial information. Spatial information today is needed for many traditional and new application fields like mobile telecommunication, electronic commerce, facility management, disaster control or environmental protection. A problem of the interdisciplinary GIS research field is that the field is still very young. Hitherto there is no common understanding about standardized GIS curricula and many GIS notions and concepts are still not precisely defined.

During the last years the GIS community tried to take the step from layer-based and map--oriented GIS to more flexible object-oriented GISs. This has not been an easy step, also because the concept of object-orientation is still not standardized in the geosciences and computer science. Geoscientists often think about structural object-orientation (DITTRICH 1986) as being the main aspect, i.e. records or object, hierarchies without concern for the behaviour of the objects. In today's GIS the data modelling of thematic, i.e. non-spatial data, is still often realized according to the Relational Data Model (CODD 1970). The "object-relational approach" combines proved relational database technology with object-oriented modelling. It is astonishing that this approach was prepared 20 years ago with the NF^2-data model (JAESCHKE and SCHEK 1981; SCHEK and SCHOLL 1986). Today's object-relational database systems, however, map the objects internally into "flat" relations. That is why spatial queries often have to be processed on more than one table and why costly "join operations" are necessary. Object-oriented database management systems (OODBMS) offer a solution for this problem. However, hitherto OODBMS basically are nothing but pure storage managers that require a lot of programming work for the "application programmer". A close cooperation between geoscientists and computer scientists is advisable, because geoscientists with appropriate software project experiences are very rarely to be found on the market. The geoscientists introduce specialized knowledge during the whole software development process and the computer scientists map the requirements into well defined interfaces and software components. This procedure leads to a permanent exchange between the geosciences and computer science.

This book intends to be a contribution to the modelling and management of inherently three-dimensional objects -particularly from geology- and their different states in time. Important concepts are topological relationships for non-primitive geo-objects and state transitions for topological relationships in time. Finally, the presented concepts are realized in the "GeoToolKit" (BALOVNEV et al. 1997a) and in a component-based GIS which is implemented on top of object-oriented CORBA technology (OMG 1997a, 1997b). The 3D geo-applications like the interactive geological modelling (SIEHL 1993) and geophysical 3D modelling show the typical requirements to 3D/4D GIS.

In this book "3D" stands for the three spatial dimensions: a "3D-GIS" is a geoinformation system that provides data structures and operations for the presentation, management and processing (analysis) of points, lines, surfaces and volumes in three-dimensional space. As "3D/4D GIS" we have in mind a geoinformation system that additionally to 3D objects is able to present and manage different states of the objects in time. In geology the temporal distance of these states often is thousands or millions of years so that the geologist has to interpolate between different states of the objects. The interpolation -under consideration of geological parameters- between the states which are stored in the database enables the visualization and the animation of a geological process like the development of basins or mountains.

In this book concrete problems in geological applications are transferred into the software development of GIS. The way from open geodatabase kernel systems to component-based GIS is presented. The component-based approach in this book is used as an instrument of software technology for the realization of modular and distributed GIS applications.

1.2 Overview

This book is organized as follows: in chapter 2 three basic "columns" of geoinformation systems are introduced: geodata modelling, geodata management and geodata analysis. Chapter 3 assesses today's commercial and prototype GIS. In chapter 4 3D/4D GIS are introduced. Especially, the data modelling and management of 3D/4D geo-objects are discussed. Common aspects and differences of spatial and temporal GIS concepts are shown. Central themes are spatial and temporal representations, temporal state transitions between topological relationships, the integrity checking and the database access for 3D/4D geo-objects. In chapter 5 the concepts of chapter 4 are applied. They are disucussed with GeoToolKit, a component-based geodatabase kernel system for the support of 3D/4D geo-applications. Furthermore, specific components for a geological 3D/4D information system are presented. In chapter 6 the integration of data and methods for distributed GIS components are examined. A case study with real geological data of the Lower Rhine Basin, Germany, is presented. Finally in chapter 7 the implementation of a distributed component-based geoinformation system is shown. The geological and the geophysical components communicate via GeoToolKit with the objective to access the shared geodatabase. Chapter 8 gives an outlook on future work.

Chapter 2

Fundamental Principles

After a short historical outline of research in the field of geoinformation science, in this chapter the particularities of 3D/4D geoinformation systems are introduced. Furthermore, the principles of component-based GIS are presented. Finally, the essential "columns" of geoinformation systems are shown and explained with examples.

2.1 "Geoinformation Science"

2.1.1 Change of the Notion 'GIS'

What Mike Goodchild claimed in his programmatic speech in Zürich at the fourth "Symposium on Spatial Data Handling" (GOODCHILD 1990) is now -10 years later- coming into vogue: he claimed that the last letter in the abbreviation *"GIS"* (for Geographical Information *System*) should be changed into *"Science"*. The universal character of the research field should be emphasized rather than special technical issues. The established journal "IJGIS[1]" followed this example and changed its name to "International Journal of Geographical Information *Science"*. Many contributions in well known international GIS conferences confirm this development.

Today it is evident that theoretical results in geoinformation science are completed with practical experiments, e.g. by new developments in software technology. Examples are the development of new user-interfaces, new system architectures or web-supporting programming languages like Java. FISHER (1998) has drawn the GIS community's attention to the valuable symbiosis between the development of Geographical Information Systems (practical aspect) and the further development of a "Geographical Information Science" (theoretical aspect). We see "Geoinformation Science" as an interdisciplinary research field being supported by computer science and the geosciences.

2.1.2 GIS Research in Germany: a Retrospective

In contrary to other (European) countries there is still no *"GIS centre"* in Germany. The GIS research has developed decentralized, mostly funded by projects of the German Research Foundation (DFG). The first and until today the most significant effort was the priority programme "Digital Geoscientific Maps" (VINKEN 1988, 1992) in which the development of

---

[1]. By Taylor and Francis, London.

prototypical geoinformation systems, e.g. for the construction of maps (GRUGELKE 1986; PREUSS 1992; GRÜNREICH 1992; VOSS and OCHMANN 1992; DIKAU 1992), 3D visualization (RAMSHORN et al. 1992), 3D modelling (SIEHL et al. 1992) and the geodata management (NEUMANN 1987; NEUMANN et al. 1992; SCHEK and WATERFELD 1986; WOLF 1989; WATERFELD and BREUNIG 1992) succeeded. At the end of the priority programme, however, the computer scientists could not yet hand over complete tools to the geoscientists. From the database side, the DASDBS geokernel, a geo-extension of a non-relational DBMS on the basis of the NF^2-data model[1] (JAESCHKE and SCHEK 1981; SCHEK and SCHOLL 1986) was designed and implemented.

Theoretical work on geo-extensions of the relational algebra (CODD 1970) and the formalism of spatial data types in 2D space have been developed by (GÜTING 1988) and by (GÜTING and SCHNEIDER 1993). (BILL and FRITSCH 1991; GÖPFERT 1991; BARTELME 1995; BILL 1996) and other authors have written text books on geoinformation systems. The intensive use of object-oriented database technology for GIS applications in the field of environmental protection has been examined at the research institute for applied knowledge processing (FAW) at Ulm (GÜNTHER and LAMBERTS 1992). The tradition of the geodatabase kernel development that started with the DASDBS geokernel (SCHEK and WATERFELD 1986; WOLF 1989; WATERFELD and BREUNIG 1992; NEUMANN 1987; NEUMANN et al. 1992) has been continued with OMS (BODE et al. 1992; BREUNIG et al. 1994) and with the development of the GeoToolKit at the collaborative research centre SFB 350 at Bonn University (BODE et al. 1994; BALOVNEV et al. 1997a,b; BALOVNEV et al. 1998a). GeoToolKit is based on an object-oriented DBMS. It was developed for the database support of 3D/4D geo-applications. Especially geological applications were consolidated in this research work (SIEHL 1993; BODE et al. 1994; ALMS et al. 1998). In the meanwhile other groups have picked up the idea of an object-oriented geo-database kernel system (BECKER et al. 1996).

The vision of "Interoperable Geoscientific Information Systems" (IOGIS) is pursued in a DFG joint project of the same name since 1996 (VOSS and MORGENSTERN 1997; OGIS et al. 1997). The objective of the project is to develop a new generation of open geoinformation systems that communicate between each other to work on a specific geoscientific goal. Important aspects are the exchange of data and methods between different GIS.

2.1.3 International Development

The synonym "GIS" for geographical information systems has been introduced in Canada by Robert F. Tomlinson (TOMLINSON 1972) in a compendium of the International Geographical Union Comission on Geographical Data Processing and Sensing. Tomlinson defined a GIS as *"not a field in itself but rather the common ground between information processing and the many fields utilizing spatial analysis techniques"*. COWEN (1988) subdivided the definitions for geographical information systems into four general approaches: *the process--oriented approach, the application approach, the toolbox approach* (see also DANGERMOND 1983) and *the database approach* (see also GOODCHILD 1985).

---

[1]. Non-First-Normal Form data model, i.e. relations need not be in the "first normal form". Therefore attributes with set-valued attributes are allowed; i.e. attributes can again contain attributes.

Today we can observe a similar development for geoinformation systems as that which took place for the hardware development some years ago. The movement from large, monolithic systems to specialized components (SZYPERSKI 1998) is obvious.

In the following historical outline we follow BREUNIG (1996). The first geographical information systems like ARC/INFO (MOREHOUSE 1985), TIGRIS (HERRING 1987), SPANS (KOLLARITS 1990) and SYSTEM9 (1992) are "historically grown software systems" whose complex user-interfaces are expecting skilled users. The GIS is distinguished from other tools like CAD systems in that the data model has realized the integration of thematic and spatial data. This means that geometry and topology are treated together with non-spatial attributes and operations. Since the middle of the eighties database experts for the first time have tried to support so called non-standard applications. They developed extensible database management systems like GENESIS (BATORY et al. 1986), DASDBS-Geokernel (SCHEK and WATERFELD 1986; PAUL et al. 1987; WOLF 1989; WATERFELD and BREUNIG 1992; NEUMANN et al. 1992), PROBE (DAYAL et al. 1987), PRIMA (HÄRDER et al. 1987), EXODUS (CAREY et al. 1988), AIM-P (LINNEMANN et al. 1988), SIRO-DBMS (ABEL 1989), GRAL (GÜTING 1989), STARBURST (HAAS and CODY 1991), POSTGRES (STONEBRAKER and KEMNITZ 1991), OMS (BODE et al. 1992). On top of these systems different applications like geo-applications have been realized. Extensible database management systems (EDBMS) enable the integration of user-defined data types -e.g. geometric data types- and of spatial access methods into the type system of the DBMS. Since the beginning of the eighties structures for the efficient management of spatial data in main and secondary memory like the quadtree (FINKEL and BENTLEY 1974; SAMET 1990), the R-Tree (GUTTMAN 1984) and the Grid File (NIEVERGELT et al. 1984) have been developed. They can today be used as components of geoinformation systems. Parallel to that further components for GIS like query languages and user interfaces have been worked out (NEUMANN 1987; EGENHOFER and FRANK 1988; BANCILHON et al. 1989; RAPER and RHIND 1990; VOISARD 1991; SVENSSON and ZHEXUE 1991; BOURSIER and MAINGUENAUD 1992). Furthermore, the development of object-relational and object-oriented prototpye systems like GEO^{++} (van OOSTEROM and VIJLBRIEF 1991; VIJLBRIEF and van OOSTEROM 1992), GeO$_2$ (DAVID et al. 1993), GODOT (GAEDE and RIEKERT 1994; EBBINGHAUS et al. 1994) and GeoStore (BODE et al. 1994) has been driven forward. The easy extensibility of GIS functions has also been considered in GRASS (GRASS 1993), THEMAK2 (GRUGELKE 1986) and SMALLWORLD GIS (SMALLWORLD 1996). For some years also object-oriented databases (ATKINSON et al. 1989; HEUER 1992; ODMG 1993) have been employed for geo and environmental applications (GÜNTHER and LAMBERTS 1992; BODE et al. 1994; BECKER et al. 1996; BALOVNEV et al. 1997a,1998a). Theoretical work mainly deals with basic concepts of GIS (BRASSEL 1993) and with the development of topological and geometric data models in 2D and 3D space (EGENHOFER 1989, 1991; PIGOT 1992a) or in spatio-temporal environments (WORBOYS 1992, 1994; PIGOT 1992b; DE HOOP et al. 1994; FRANK 1994; HEALEY and WAUGH 1994; PILOUK et al. 1994; VAN OOSTEROM et al. 1994).

Today in the geosciences more and more models are required for the exchange and the integration of spatial information (GUENTHER and BUCHMANN 1990; WORBOYS and DEEN 1991; ABEL et a. 1992; GÜNTHER and LAMBERTS 1992; SALGE et al. 1992; SCHEK and WOLF 1992, 1993; WOLF et al. 1994; BREUNIG 1996; BUEHLER and MC KEE 1996; GREVE et al. 1997). This request for data integration is reinforced by the tendency of geoscientific disciplines to understand themselves increasingly as environmental sciences. From a computer science point of view data integration is particularly interesting, if information is managed and processed by spatially distributed systems or even on heterogeneous hardware platforms.

2.1.4 3D/4D Geoinformation Systems

The development of 3D/4D geoinformation systems is still at the beginning (RAPER 1989; TURNER 1992; GOCAD 1996; ALMS et al. 1998). Theoretical approaches for 3D geoinformation systems preferentially deal with topological data models (DE HOOP et al. 1994; PILOUK et al. 1994). Many contributions in GIS research tried to solve the problem of modelling and managing 3D data (BURNS 1975; VINKEN 1988, 1992; TURNER 1992; KELK 1992; ABEL and OOI 1993; BRUZZONE et al. 1993; DE FLORIANI et al. 1994; HACK and SIDES 1994; HEALEY and WAUGH 1994; PILOUK et al. 1994; VAN OOSTEROM et al. 1994; BODE et al. 1994; BREUNIG et al. 1994; BREUNIG 1996; BALOVNEV et al. 1997a; ALMS et al. 1998). In addition, DIGMAP (DANN and SCHULTE-ONTROPP 1989; GRAPE (1997) and GOCAD (MALLET 1992a; GOCAD 1996, 1999) can be denoted as components of a 3D geoinformation system. Temporal data handling has been examined by (SNODGRASS 1987; WORBOYS 1992, 1994; SNODGRASS 1995; VOIGTMANN et al. 1996; BILL 1997; SPACCAPIETRA et al. 1998), and by other authors.

A 3D/4D geoinformation system distinguishes from other software systems as follows:

1. The 3D/4D data modelling (for objects in three-dimensional space and their state transitions in time);
2. Spatial and temporal management of geodata (with spatio-temporal access methods);
3. The 3D/4D visualization (for the animation of geoscientific processes).

In the data handling module of today's commercial 2D geoinformation systems the third spatial dimension is at best dealt with as a thematic attribute like the height value of an isoline map. It is confusing that in geography "3D" is often interpreted as "2D plus time", i.e. as the management of two-dimensional geometries in different time steps. Unfortunately, vendors of geoinformation systems often also call the dealing with an arbitrary thematic attribute of a two-dimensional geometry as "3D representation". Mostly the data handling of today's GIS is not coupled with the data types used for visualization. Furthermore, the spatial data handling in today's GIS is in most cases not executed by a database management system, but directly by the operating system. However, it should be the objective for future generations of GIS to manage complex 3D/4D objects within their data handling component.

2.1.5 On the Way to Component-Based Geoinformation Systems

The development away from centralized mainframe configurations to distributed client/server architectures means for future geoinformation systems that data and methods have to be controlled from outside the system. The opening of the geoinformation system for external software tools will have the advantage for the geoscientist to have a shared access on different data sources and geo-services. This simplifies the dealing with interdisciplinary problems. However, there is also the danger of global network systems becoming too complex to be handled efficiently. Furthermore, the compatibility of data and services has to be checked for each application separately. It is known that until today there are security problems for the access of world wide connected software systems and it is also understandable that the free interdisciplinary access and evaluation of personally acquired data is not desirable in every case (copyright problem). Taking it all in all, we should see the benefits for the geosciences gained from the integration of heterogeneous data and software systems. We now state more precisely the notion of an *open geoinformation system*. We require the following conditions for the openess[1]:

1. The externally visible functionality of the geoinformation system has to be extensible and changeable;
2. The geoinformation system must have a well documented interface;
3. There are system levels that should not be open. For example, the internal data structures of the geoinformation system should not be changeable from outside the system.

The conditions above make clear that openess is mainly a point for the developer of geoinformation systems and not for the geoscientific end-user. But the change away from classical programming interfaces to easily changeable building-block like software environments increasingly helps the geoscientist to adapt software systems for special needs.

The necessary data conversion between the special data formats of formerly closed geoinformation systems causes high costs. A first trial to put data sharing and interoperable use of geoinformation systems into practice is the OpenGIS consortium. It was founded in 1994 by established geoinformation system, database and hardware vendors. Also some universities like UC Berkeley, University of Münster, TU Vienna and University of Zürich are members. In the "Open Geodata Interoperability Specification" (OGIS) (BUEHLER and MC KEE 1996) the specification of an "OGIS-Framework" is proposed which consists of an open 3-level-model for geodata, services and information sharing. This abstract specification was developed to ensure the independence of OGIS implementations like CORBA, OLE, Java. The "Open Geodata Model" (OGM) serves as an abstract model for the digital representation of the Earth and its phemonena. It is similar to the object, dynamic and functional models of RUMBAUGH et al. (1991). The "OGIS Services Model" is thought of as specification model to implement services according to a client/server architecture for the geodata access, the manipulation, representation of geodata as well as the sharing of so called information communities[2]. Part of the model is a query model that contains an object query language (OQL) near to the ODMG standard (ODMG 1993). This query language has been especially extended for spatial and temporal boolean predicates (*intersect, contain, equal* etc.) and for spatial object types *(point, curve, surface, solid etc.)*. Finally the "Information Communities Model" serves as a framework for the solution of technical interoperability problems during the use of the Open Geodata Model and the OGIS Services Model. The specification language follows the Interface Definition Language (IDL) of the Object Management Group (OMG).

---

[1] The following points have been proposed during a discussion of researchers working in the IOGIS-project (VOSS and MORGENSTERN 1997).
[2] In the OGIS context an information community is the union of a collection of geodata and services.

The extensibility and opening of geoinformation systems is the first step on the way to interoperability between hitherto isolated components. With extensibility we mean the property of a software architecture to support unforeseen requirements and to adapt the system according to these new requirements. By interoperability we understand the capability to exchange functionality and interpretable data between software entities. For that a communication channel and a general communication protocol are needed. Across the channel the entitities have to formulate and transfer interpretable queries to the functions or to the data. The result of the query then has to be sent back in an intermediate data format. This also includes a syntactic and semantic checking of the query. Thus interoperable geoinformation systems are "more" than open geoinformation systems. They communicate with each other to solve a geoscientific problem. Precondition is that each geoinformation system in the network shall also be working independently as an autonomous local system. We require the following conditions for *interoperable geoinformation systems:*

1. Each interoperable geoinformation system must autonomously be operable and it must be able to communicate with at least one other geoinformation system in the network;
2. The objects exchange (data and methods) between interoperable geoinformation systems must be provided via a well defined and documented interface;
3. For each geoinformation system G in the network there is a geoinformation system G', that can replace G so that the systems network as a whole keeps operable.

Interoperable GIS should be component-based systems. The component-based architecture simplifies the extension of the system, the adaptation of new requirements and the maintenance of the system. The point is to modularize the requirements, architectures, design and implementation of software systems (SZYPERSKI 1998). The components must interact with each other and they have to build an operable network. It should be mentioned that the single components depend on each other. There is an obvious benefit from using them together. From an economic point of view component software becomes more important, because out-sourcing of software development is increasing. However, there is no significant experience with component-based GIS technology. This book tries to make a contribution to this area.

2.2 Geodata Modelling

2.2.1 Peculiarities of Geoscientific Data: Spatial and Temporal Reference

BILL and FRITSCH (1991) mentioned that data has the longest life of the three "GIS columns" hardware (3-5 years), software (7-15 years) and data (25-70 years!). Geodata are distinguished from other data by their spatial and temporal references. The *spatial reference* gives information about the absolute and the relative spatial position of geo-objects[1] or about observed geoscientific phenomena like the absolute Gauß-Krüger coordinates of a volcano, its neighbouring objects or volcanos near to the observed one. Geo-objects have a spatial extension, if they are non-point objects. They often spread over large areas of several kilometres. This is not the case with CAD objects.

---

[1] The notion of the geo-object is more closely looked at in chapter 2.1.2.2.

The *temporal reference* gives information about the absolute and relative temporal position. Examples are times or time intervals in which different states of a geoscientific phenomenon are stored in the database. In future the evaluation of spatial and temporal geodata could enable geoinformation technology to support the solution of world wide problems like environmental monitoring. The high benefit of geoinformation systems for the observation of thunderstorms and high waters or for the computation of the shortest path in train networks is indisputable. Furthermore, geoinformation systems are well suited for the management of infrastructural data like electricity, gas or telephone mains. The GIS work well in these fields, however, new requirements are not always easy to meet so that GIS developers do well to revoke the high expectations of GIS users.

2.2.1.1 Pointset Topology

The geometry of geo-objects can be described abstractly with point sets which enables an abstraction from the concrete geometric realization.

Topological relationships on point sets like equality, inclusion and intersection can be defined with the well known set operators. We give some examples (see also GÜTING 1988). Let *a, b* be geo-objects and let *geom(a)* and *geom(b)* be the geometry of *a* and *b*, respectively. Then the following topological relationships are defined on point sets:

geom(a)	= *geom(b)*	:= $points(a) = points(b)$;
geom(a)	≠ *geom(b)*	:= $points(a) \neq points(b)$;
geom(a)	inside *geom(b)*	:= $points(a) \subseteq points(b)$;
geom(a)	outside *geom(b)*	:= $points(a) \cap points(b) = \emptyset$;
geom(a)	intersects *geom(b)*	:= $points(a) \cap points(b) \neq \emptyset$.

Metric relationships between point sets (distances) can be defined on labelled points like the centre of gravity.

With the *topology* of a geo-object we mean the connection of nodes, edges, surfaces and volumes of a geometry being approximated from the real world geometry. The approximated *geometry* of a geo-object is the set of its coordinates and the mathematical equations of its segment, surface and volume elements, respectively. The topology of a triangle network for example, is the triangle surfaces ("neighboured triangles") which are connected by edges. They are often numerated by integer numbers. The geometry of the triangle network consists of the coordinates of the base points of all triangles building the plane equations of the triangles.

2.2.1.2 Algebraic Topology

Algebraic topology provides an approximated computer representation for the abstract geometric data types *point, line, surface and volume* with the concept of simplicial complexes.

Definition 2.1: The points $p_0, p_1, ... p_k$ are called *affinely independent*, if the k-tuple ($p_0p_1, p_0p_2, ... p_0p_k$) is linearly independent.

The p_0p_i in definition 2.1 are elements of the vector space of translations. They can be seen as free vectors that are "attached" to the points.

Definition 2.2: Let P be a set of (d+1) affin independent points.

Let *conv(P)* be the convex hull of P and $P' \subset P$. Then $S := conv(P)$ is called a *d-dimensional simplex* and *conv(P')* is a *face* of S. The set of all faces of a simplex is called *face(S)*.

A set C of simplexes is called *simplicial complex*, if
1. For each $s \in K$: $face(s) \subseteq C$
2. $s1 \cap s2 \neq \emptyset \rightarrow s1 \cap s2 \in C$

The dimension of a simplicial complex C $dim(C) = max\{dim(s) \mid s \in C\}$

In n-dimensional space an n-simplex has $\binom{n+1}{d+1}$ faces of dimension d. For example, a 3--simplex in 3-dimensional space has $\binom{3+1}{0+1}$, i.e. 4 points, $\binom{3+1}{1+1}$, i.e. 6 edges, $\binom{3+1}{2+1}$, i.e. 4 triangles and $\binom{3+1}{3+1}$, i.e. 1 tetrahedron as faces.

If C is a simplicial complex of dimension d $(d > 0)$, then the boundary of C $(@C)$ is a simplicial complex of dimension $(d - 1)$.

Obviously, simplexes have a simple structure. In each dimension they consist of the simplest possible geometry, respectively.

The precise definition of spatial relationships between geo-objects is the theoretical basis for the understanding of spatial database queries. The theory of topological relationships based on pointset topology and the application of topological relationships for the definition of spatial database queries in GIS has been examined in detail by (EGENHOFER 1989; EGENHOFER and FRANZOSA 1991, 1995; PIGOT 1992a. As EGENHOFER (1989) proposed, simplexes can be used well for the realization of binary topological relationships between geo-objects. Topological relationships are based on the concept of topological space and on the notions *"interior"*, *"hull"* and *"boundary"* of open sets. The intersection of the interior with the boundary of an open set is empty and the union of the interior with the boundary is the hull of the set.

The so called *4-intersection invariant* of EGENHOFER and FRANZOSA (1991) distinguishes between empty and non-empty intersections of the boundary and the interior of regions. It is the basis for the definition of a *minimal set of binary topological relationships* between lines only and areas only in two-dimensional space. The relevant 6 topological relationships are *disjoint, contains/inside, meet, equal, covers/coveredBy* and *overlap* (see Fig. 2.2). Precondition is the existence of elementary boundary and interior operations.

The *boundary operator* returns a set of (k-1)-simplexes for a k-simplex. To complete the operations we also list the *coboundary operator*, which returns the neighboured (k+1)-simplexes of the k-simplex. Figure 2.1 shows the boundary and the coboundary operator for a 2-simplex. The result of the operator is drawn bold, respectively.

Chapter 2. Fundamental Principles   11

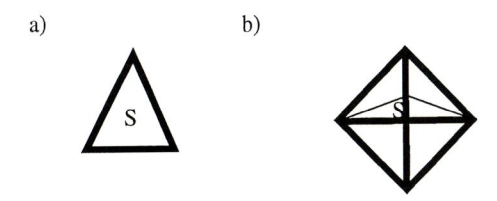

a)   b)

Fig. 2.1 a) Boundary of a 2-simplex S;
b) Coboundary of a 2-simplex S

In Fig. 2.1a) the result consists of the three 1-simplexes of the triangle. In Fig. 2.1b the result is the two adjacent tetrahedra (3-simplexes).

disjoint   meet   overlap

covers/coveredBy   inside/contains   equal

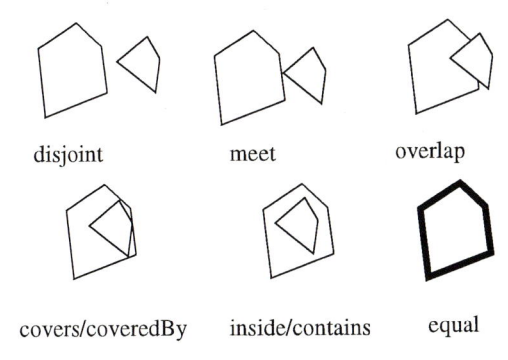

Fig. 2.2 Minimal set of topological relationships between simplicial 2-complexes by Egenhofer

Figure 2.2 and Table 2.1 show the 4-intersection invariant of Egenhofer and Franzosa with the four possible intersections (in 2-dimensional space) of the boundaries and the interiors of two simplicial complexes C_1 and C_2 (@$C_1 \cap$ @C_2, °$C_1 \cap$ °C_2, °$C_1 \cap$ @C_2, @$C_1 \cap$ °C_2, with @ ≡ boundary, ° ≡ interior). The specification of the topological relationships for simplicial n-complexes (n > 0) is shown in Table 2.1. The criterion for the intersection tests are empty (∅) and non-empty (¬∅) intersections.

C_1 C_2	$@\cap @$	$^o\cap^o$	$@\cap^o$	$^o\cap @$
disjoint	$\varnothing$	$\varnothing$	$\varnothing$	$\varnothing$
meet	$\neg\varnothing$	$\varnothing$	$\varnothing$	$\varnothing$
overlap	$\neg\varnothing$	$\neg\varnothing$	$\neg\varnothing$	$\neg\varnothing$
covers	$\neg\varnothing$	$\neg\varnothing$	$\varnothing$	$\neg\varnothing$
coveredBy	$\neg\varnothing$	$\neg\varnothing$	$\neg\varnothing$	$\varnothing$
inside	$\varnothing$	$\neg\varnothing$	$\neg\varnothing$	$\varnothing$
contains	$\varnothing$	$\neg\varnothing$	$\varnothing$	$\neg\varnothing$
equal	$\neg\varnothing$	$\neg\varnothing$	$\varnothing$	$\varnothing$

Table 2.1 The specification of a minimal set of topological relationships based on the criterion of empty and non-empty intersections of boundaries and interiors.

The *overlap*-relationship returns the value "true", if the boundaries of both complexes intersect themselves ($@\cap @\equiv\neg\varnothing$), if boundary and interior intersect ($@\cap^o\equiv\neg\varnothing$) and ($^o\cap @\equiv\neg\varnothing$), and if the interiors of the complexes intersect each other ($^o\cap^o\equiv\neg\varnothing$). *Covers* and *coveredBy* as well as *inside* and *contains* are redundant relationships, i.e. *covers* can be simulated by *coveredBy* (or vice versa). *Inside* can be simulated by *contains* (or vice versa) by changing the order of the operators. That is why the *coveredBy*- and the *contains*-relationship can be deleted from Table 2.1.

The 9-intersection invariant (EGENHOFER and FRANZOSA 1991) has been introduced to describe relationships between lines, and between areas and lines. It additionally includes intersections between the exterior of regions. The 4-intersection invariant has been extended by CLEMENTINI and DI FELICE (1994) for topological relationships in $\Re^3$. The extension facilitates the intersection between finer topological configurations like the intersection between two objects in one plane and two objects in two different planes. Additionally, the dimension for the geometries of the intersection result is used in this approach. EGENHOFER and FRANZOSA (1995) have further refined the 4-intersection invariant by adding several parameters. The resulting invariant distinguishes between topological relationships of concave areas and it even considers the number of connected objects that are generated during the intersection.

2.2.1.3 Pointset and Algebraic Topology in Time

Naturally, phenomena in the geosciences can be observed in space and time. Comparable to space we can also distinguish in time between an absolute and a relative time. Following WORBOYS (1995) we introduce an example for the so called travel time topology that can be seen as a topological space. Let P be the point set that geographically covers the Rhine-Neckar region near Heidelberg in Germany. In the region a public travel network exists. Let the average travel time be known between two arbitrary points of the network. Furthermore, let the travel time be symmetric, i.e. one gets exactly as fast from point *A* to *B* as from point *B* to *A*. Under these conditions the travel time topology is a topological space (WORBOYS 1995). It can easily be shown that the two following conditions hold:

1. Every point has a time zone that surrounds the point;
2. The intersection of two arbitrary time zones (corresponding to the neighbourhood in space) of an arbitrary point in the region of the travel network contains a time zone of that point.

As we have seen in the example above, metrics can be defined on time zones that meet identity, symmetry and the triangle inequality. Euclidean metrics are a special case of such metrics.

PIGOT (1992a,b) has defined spatio-temporal objects based on cell complexes in a 4D Euclidean space. In this approach time is a further spatial dimension. The objects are represented as (k+1)-manifolds (k spatial and one temporal dimension) in 4D Euclidean space. I.e. processes are seen as a sequence (snapshots) of k-dimensional well defined cell complexes in time. The cell complex is a generalisation of the simplicial complex and it is based on the abstract notion of cells. One of the preconditions in Pigot's model is that the connectivity of the cells belonging to a cell must not be changed between two snapshots, i.e. the topology of two neighbouring states is equal. The advantage of this approach is that well known geometric operations like the set operations (intersection, union, difference) can simply be extended to $\Re^4$.

2.2.2 Approaches of Spatial Data Models

A model abstractly describes phenomena of the real world. It is independent of the implementation of the model. A map is a well known geoscientific example for a model. It shows a person the way, for example from the city of Heidelberg to the city of Bonn. During the development of geoinformation systems two approaches for spatial data models have been established: the field-based models and the object-based models (see also WORBOYS 1995). Behind these models two inverse paradigms appear. They are also realized in today's GIS as the so called raster and vector representation. Field-based models describe spatial distributions as functions of a spatial partition like a grid or TIN on an attribute domain. Contrary to that in object-based models, uniquely identificable objects with a spatial reference are in the centre of interest.

2.2.2.1 Field-Based Models

In the field-based approach geoscientific information is attached with spatial fields. Every field describes an instance of an attribute's spatial distrubution as a function of a set of locations on a finite attribute domain.

A *field* is a function F so that $F : \{R_i \mid R_i \subset R\} =: T \rightarrow D_A$

Let R be the corresponding region of space, T the tesselation and D_A the domain of the intererested geo-attribute A, then the following three conditions hold:

1. $|T| = n \in N$

2. $\cup_{i=1}^{n} R_i = R$ and $R_i \cap R_j = $ boundary$(R_i) \cap $ boundary(R_j) for all $i \neq j$

3. For all $p \in R$ there is a $t \in T$ so that $p \in t$

The three conditions above correspond to a tesselation of the region, i.e. a non-overlapping (2) and complete (3) partition of space. To every cell R_i a field value is assigned. Typical examples for such geo-attributes are the deposit or the height of a cell in a digital elevation model (DEM). In a DEM the digitally stored height values of the Earth's surface are described as a function of their distance from the Euclidean plane (z-values). The only attribute of a DEM is the height, i.e. the distance to the (x,y)-plane.

For a spatial object that is represented as a DEM the following condition is held: for every (x,y)-value only one z-value is allowed, i.e. $z = f(x,y)$. Thus "overhangs" in terrain cannot be represented in a DEM. We speak of a "2.5-dimensional surface".

A spatial object O is called 2.5-dimensional, if it is embedded into Euclidean space $\Re^3$ and if it has the following property: if $(x,y,z) \in O$ and $(x,y,z') \in O$, then $z = z'$.

In practice DEM are often represented as a regular grid (elevation matrix) or as a triangulated irregular network (TIN). The advantages of a regular grid are the simplicity of the modelling and the availability of many algorithms for spatial analysis on grids. However, it is less flexible in the scale, because its resolution is the same at every location. Thus the grid cannot be adapted to special structures of the landform. A TIN does not have this disadvantage, but spatial analysis is more complex because of its irregular structure. As an example one could determine the amount of deposit in a given geographical region like Heidelberg and interpolate the values of the existing measuring instruments on a regular grid. A more detailed model could be obtained by using a TIN with the location of the measuring instruments as points of the TIN to be interpolated.

For several years hierarchical multi level methods have been used for the storage and representation of DEM (BRUZZONE et al. 1993; GRIEBEL 1994). In Fig. 2.3 three operations are given that typically can be applied to a DEM.

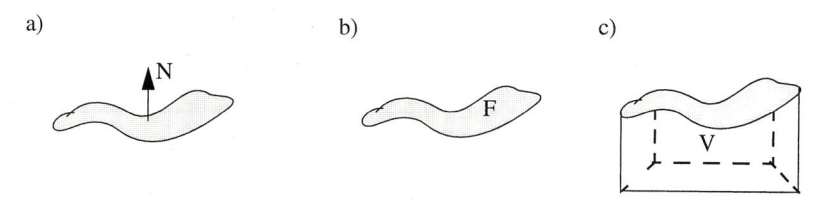

Fig. 2.3 Operations applied to a DEM:
 a) Normal vector of a point of the Earth's surface, e.g. for the determination of the slope gradient
 b) Area of the Earth's surface within a given region
 c) Volume under the Earth's surface from a given depth and region

With spatial field functions we can map spatial fields to attribute values. So GIS analysis can be executed on top of the field functions; for example, the linear interpolation of the deposit with a TIN derived from point-based measurements. A field function has as input one or more fields that can be either spatially equal and thematically different (neighbouring fields) or spatially different and thematically equal like different thematic layers of a GIS. The field function again returns a field as its result.

2.2.2.2 Object-Based Models

In the object-based models flexible objects in space like cities, countries, rivers etc. with their attributes like the number of inhabitants, size, length etc. are considered and not spatial distributions on strictly given fields. Attributes can either statically be stored and retrieved in the database or they can be computed on-demand by the database. An example is the computation of the average number of inhabitants of all cities in North Rhine Westphalia, Germany. On the one hand this approach corresponds with object-oriented modelling, on the other hand it can also serve as a framework to describe field-based and object-based models.

In the object-oriented paradigm a geo-object class can be modelled in two ways. We distinguish between the thematically dominant and the spatially dominant geo-object modelling (Fig. 2.4). In the thematically dominant geo-object modelling a thematical geo-object class is defined. Geometry and topology are defined as an attribute or operation besides the thematical information and the graphical description. The graphical description serves as the graphical representation of the geo-objects, e.g. on a map. The behaviour of a geo-object can contain methods like the visualization of the geo-object within a certain scale or the distance between two geo-objects.

```
 a)  ┌─────────────────────────┐    b)   ┌─────────────────────────┐
     │    geo-object class     │         │   spatial object class  │
     │      (thematical)       │         │      (geom/topol)       │
     ├─────────────────────────┤         └───────────△─────────────┘
     │ thematical information  │                     │
     │ spatial description     │                     │
     │ geometry/topology       │         ┌─────────────────────────┐
     │ graphical description   │         │    geo-object class     │
     ├─────────────────────────┤         │     (geo-specific)      │
     │       operations        │         └─────────────────────────┘
     └─────────────────────────┘
```

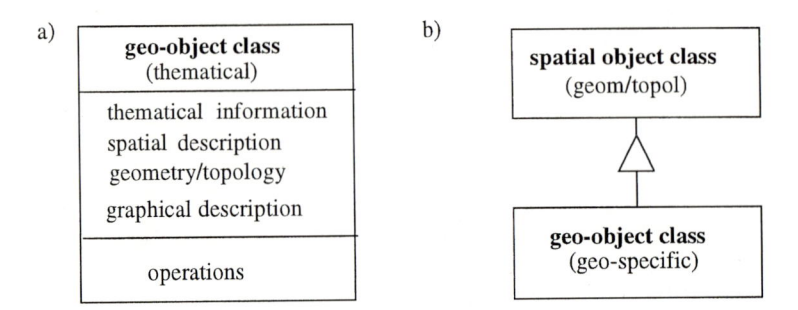

Fig. 2.4 Two ways for the modelling of geo-object classes
  a) thematically dominant modelling
  b) spatially dominant modelling

In the spatially dominant geo-object modelling the thematical geo-object class inherits the funtionality of the spatial class, i.e. attributes and operations. It adds its own specific functionality.

Today, both models (field-based and object-based) have equal rights. They are often used together in the so called hybrid approach.

Hierarchy of Spatial Object Classes

In many geo-applications hierarchically organized spatial object classes play a dominant role. Following (WORBOYS 1995) we describe the principle of spatial object classes.

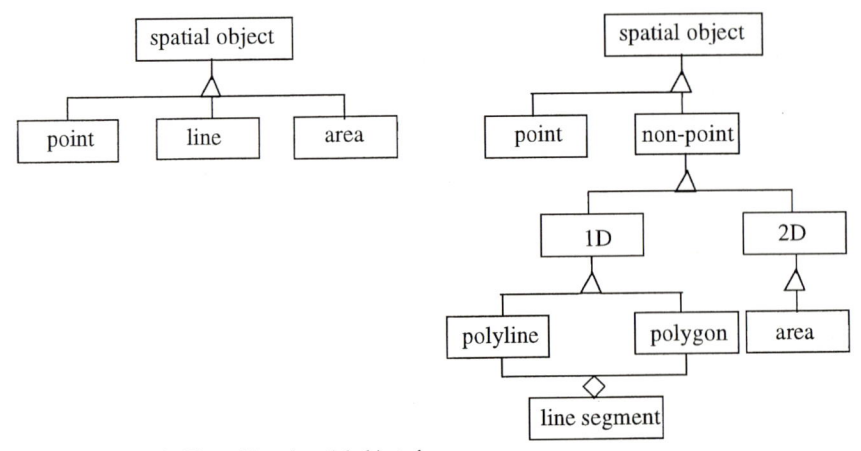

Fig. 2.5 Examples for hierarchies of spatial object classes

Figure 2.5 shows two examples for hierarchies of spatial object classes in the Euclidean plane. The specialization that starts from a general spatial class *spatial object* and branches to point, line and area classes provides a flexible extension of the class hierarchy. Special geometries inherit the functionality of the *spatial object* class and they add their own attributes and operations.

2.2.3 Temporal Extension of the Field-Based and the Object-Based Approach

Similar to space we can also distinguish in time between the field-based and the object-based approach (see also WORBOYS 1995), i.e. we can distinguish between temporal distributtions and between objects with a temporal attribute or an object with a time stamp.

On the one hand geoscientific processes like the movement of a thunderstorm in space and time can be expressed with spatio-temporal field functions from which each of them describes the change of a measured attribute in space and time. This corresponds to the representation of a scene that changes in time. On the other hand such a process can be described with single spatio-temporal objects. Every object can independently be modelled with its own time, i.e. parts of the process can spatially and temporally be examined in more detail. We will again pick up the object-based approach to time in chapter 5.4.5.

2.2.4 Implementation of Spatial Data Models

2.2.4.1 Vector Representation

In today's GIS the vector representation is used as the implementation of the object-based model. This representation is also called "vector data model" (PEUQUET 1984) or "vector format" (WORBOYS and DEEN 1991).

<u>Advantages and Disadvantages of the Vector Representation</u>

The advantages of the vector representation are its high precision, the small storage space needed, the simple execution of coordinate transformations like the addition of the coordinates by a constant value in x-, y- and z-direction and the simple realization of distance operations like the Euclidean distance.

A disadvantage of the vector representation is that the intersection of areas and the determination of neighbourhoods are difficult to execute and too costly in the general case (MEIER 1986)[1].

2.2.4.2 Raster Representation

The typical implementation of the field-based model in today's GIS is the raster representation. More generally we speak of tessellated data structures (RHIND and GREEN 1988), because they divide the plane into discrete cells. These data structures are also called tessellation models. The procedure is opposite to that of the vector representation: geometries are not built from points and referred to objects, but the plane is decomposed into independent cells. In the vector representation the data modelling starts with "spatial objects", whereas

---

[1]. On average $O(n^2)$ corresponding to the number n of the involved lines.

the tessellation models describe the spatial continuum. However, a reference from space to single geometries is also possible with tessellation models. An example is the classification of raster cells. Tessellated representations can be classified as follows (see RHIND and GREEN 1988):

1. Regular tessellations (e.g. grid or raster);
   - nested regular tessellations (e.g. quadtrees);

2. Irregular tessellations (e.g. triangulated irregular networks TIN or voronoi diagrams);

3. Nested irregular tessellations (e.g. point quadtrees, k-d trees).

The partitions of regular tessellations are equal in size and they consist of regular polygons. The most known tessellations are square, triangle and hexagon. Regular tessellations are "absolutely" fixed a priori, whereas irregular tessellations are "data dependent", i.e. the tessellations result from the partition and from the order in which data is inserted.

The raster representation is traditionally the most used tessellation representation in the geosciences. It is also described as "raster model". Figure 2.6 shows the boundary of a map (coarse sketch) of North Rhine Westphalia, Germany in the vector and raster representation, respectively.

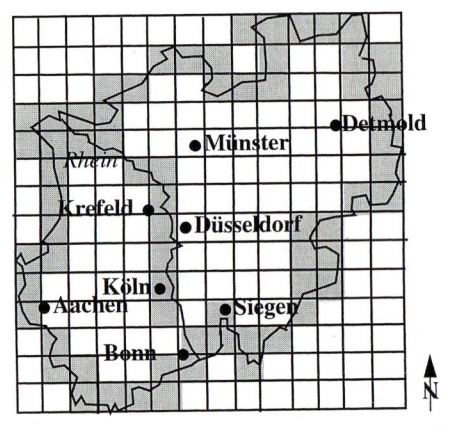

Fig. 2.6 Map of North Rhine Westphalia, Germany (coarse sketch) in the vector and in the raster representation

The notion of a *raster* is often used for two different reasons. On the one hand by raster we mean a grid which is interpolated from measure points to an area. Let us call such a grid *"georaster"*. To each raster cell a set of thematical attributes like "soil use", "deposit" etc. is assigned. From the georaster and its single raster cells geometric objects can be generated with classification methods like component labelling (see DIKAU 1992). On the other hand, the raster is known from computer graphics. The number of raster cells (pixels) determines the resolution of a picture. A cell consists of shades of grey, e.g. of a satellite picture, which determines the resolution of the picture. Let us call this kind of raster *"graphics raster"*. If we reduce the size of the cells of a georaster against zero and if we replace the thematic attributes with shades of grey values, then the georaster (grid) turns into a graphics raster.

The quadtree representation (FINKEL and BENTLEY 1974) is a refining of the raster representation with reduced storage space. Space is subdivided into four quadrants. This results in a fourtimes subdivided and tree-like data structure. The quadtree provides efficient spatial access.

Advantages and Disadvantages of the Raster Representation

The advantages of the raster representation are the disadvantages of the vector representation and vice versa. The topological properties like order, neighbourhood or connectivity properties are only given implicitly by the neighbourhood of the single cells. It is straight forward to compute the sum and the difference of elementary regions with rasters, as the raster operations on single cells can be implemented by simple bitwise operations. However, rotation and coordinate transformations are costly. The spatial resolution of the raster is dependent on the smallest representable cell. The storage space needed for a raster can be reduced with compression techniques. An example of a raster GIS that provides 3D functionality at least for visualizing objects is ERDAS IMAGINE and the IMAGINE Virtual GIS that supports real time fly over, i.e. a 2.5-dimensional view of landforms from different viewpoints.

2.2.5 Layer Model

In most of today's available GIS geo-objects cannot be defined flexibly, but they are connected with so called layers, i.e. map sheets. The advantage of this solution is that the intersection of the layers can easily be computed. For example, the intersection of different maps for inshore waters, electricity lines, soil use and average deposit in the region of Heidelberg results in a new map which considers all four aspects in one map of the GIS (see Fig. 2.7). The integration of spatial and non-spatial information is handled with an explicit pointer mechanism which is visible for the user. It combines the geometries and topologies of a map with one or more relations (tables) and their thematic attributes (see Fig. 2.7). Thus it is possible to follow the map from the table and vice versa.

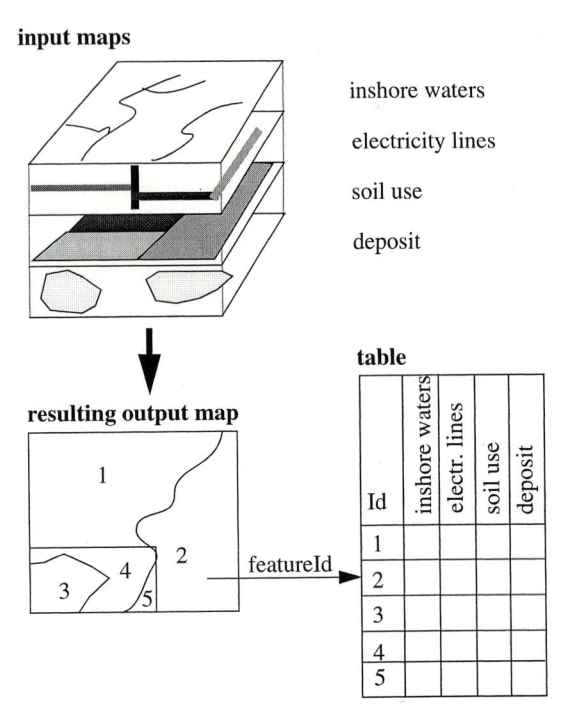

Fig. 2.7 Intersection of different map sheets and reference to the non-spatial attribute data in a GIS

In principle a consequent extension of the layer concept for 3D space, i.e. the intersection of different thematically defined volumes or even 4D layers (animation of different time-dependent volumes) is possible. However, such techniques are not yet applied to today's GIS.

2.2.6 GIS Types

The two paradigms of the field-based and the object-based approach have essentially influenced the development of geoinformation systems until today. Corresponding to their implementations we can distinguish between raster and vector GIS. The so called hybrid GIS can process raster as well as vector data. Finally, the 3D/4D GIS are a new type of geoinformation systems that will gain in importance.

2.2.6.1 Vector-Based GIS

A vector-based GIS has the following characteristic properties:

1. The vector data reach the GIS by digitalization of the data with a tablet or by using input routines from external files;
2. It supports data handling of spatial and non-spatial attributes of geo-objects;

Chapter 2. Fundamental Principles 21

3. It contains data interfaces for the conversion of input raster into vector data;
4. Resulting GIS data can again be converted into the raster format (this can lead to a loss of information);
5. It provides typical vector-based GIS functions like the distance between two objects, the intersection of adjacent areas on layers of a map or the buffer construction around point, line and area objects. A typical example is the marking of regions that are threatened by floods of a river.

Many known GIS like ARC/INFO (SCHALLER 1988), SICAD (SONNE 1988), SYSTEM 9 (SYSTEM9 1992), ATLAS-GIS etc. are predominantly oriented to the processing of vector data. Additionally, in these GIS raster data like satellite pictures can be used as the background of vector maps. An example is the combination of a digitalized vector traffic plan with a topographic raster map. However, these *hybrid GIS* usually do not automatically combine single raster cells to "objects". That is why the intersection of raster and vector data cannot be computed straightforwardly. The functionality provided for integrated data handling of raster and vector data is still limited in hybrid GIS.

2.2.6.2 Raster-Based GIS

The following properties are typical for raster-based GIS:

1. The data reach the GIS by scanning or reading the raster data from a file;
2. It supports the data handling of thematic attributes on single raster cells with a database mangement system;
3. The spatial data handling and processing is particularly oriented to raster data, i.e. the GIS supports efficient raster algorithms;
4. It contains data interfaces for the conversion of input vector data into raster data;
5. The GIS results can again be converted into the vector format.

Commercially available raster-based GIS like SPANS (KOLLARITS 1990), GRASS (GRASS 1993), ERDAS, IDRISI, SMALLWORLD GIS (SMALLWORLD 1996) etc. often originated from image processing systems. They usually contain less advanced functionality for the processing of vector data.

2.2.6.3 3D/4D Geoinformation Systems

Today's geoinformation systems are still mainly 2D information systems. We can subdivide the first approaches for 3D or 4D GIS into two categories:

1. The 2D GIS with 3D or 2.5D extensions;
2. The 3D/4D modelling and visualization tools, e.g. for geology.

The systems of the first category support the visualization of TINS or grids for digital elevation models (2.5D). Some GIS also allow the visualization of 3D scenes. To the knowledge of the author, however, advanced geometric 3D operations are not yet provided.

The systems of the second category are advanced 3D/4D modelling and visualization tools that support many algorithms to analyse 3D geometries. Examples are specialized interpolation algorithms (MALLET 1992b). These tools are able to visualize movements of 3D objects. Examples of such systems are GOCAD (MALLET 1992a), GRAPE (1997), Geom3View (PFLUG et al. 1992) and LYNX (1996). GRAPE supports a sophisticated time model for the simulation of 3D geometries and topologies in time (POLTHIER and RUMPF 1995).

Hitherto there are no standards known for 4D representations of geo-objects in GIS. However, the OpenGIS consortium (BUEHLER and MC KEE 1996) and other national standardization committees started an initiative to standardize geo-objects. During the last years cell representations have become popular for the representation of surfaces and solids (MALLET 1992a; ALMS et al. 1994; CONREAUX et al. 1998; GOCAD 1999), because they are very useful for the visualization of irregular geo-objects. For the representation of digital elevation models hierarchical tesselations have been developed (DE FLORIANI et al. 1994). The introduced systems of the first and second category mentioned, however, do not support an efficient handling of spatial and temporal data in a database management system.

2.3 Geodata Management

Besides the geodata modelling also the geodata management of large sets of spatial and temporal data in a DBMS and the geodata analysis are also central tasks of a GIS.

2.3.1 Functionality of a Geodatabase

On the one hand a geodatabase should support the functionality of a customary database management system (DBMS) which has been the practice in companies for many years. The concurrency control, crash recovery, automatic integrity control of data as well as the mechanisms to guarantee data security are central tasks of a DBMS. On the other hand a geodatabase should support geometric data types in their data model and in their query language (GÜTING 1994). Spatial relationships, functions and operators between geo-objects should be provided by extended geometric data types. Furthermore, the geometric data types should be supported in the implementation of the DBMS with spatial access methods and efficient geometric algorithms.

Relational DBMSs -which are the standard technology today- support the shared access of many users to one common database. Application and system errors are tolerated in a limited way. The *GIS transactions*, however, like the construction of a land utilization map can last several days or weeks so that the concept of short transactions does not help. It is not acceptable to lock the corresponding objects for such a long time to all other users. Thus the known "ACID" concept (**A**tomarity, **C**onsistency, **I**solation, **D**urability) has to be extended for GIS applications. According to the ACID concept an operating programme or object on top of the database is seen as one logical unit. This means that during its execution the database is exclusively available for this programme or object (atomarity). The consistency of the database is retained and any effects that are followed by the execution of the transaction must not be threatened by hardware or software errors.

Chapter 2. Fundamental Principles   23

Unfortunately, the well known "2-phase locking protocol" does not support long GIS transactions. This protocol assumes that no lock can be announced before the last lock is unlocked. Thus different object variants which have been created during a GIS session would be overwritten by this locking mechanism.

The GIS like Smallworld GIS (SMALLWORLD 1996) and some OODBMS like ObjectStore (version 3.0) support multi versioning of objects. Every write operation creates a new version of an object. The handling of the different versions, however, is the responsibility of the user.

One of the most important database functions for GIS is the efficient spatial access on a set of persistently stored geo-objects. In the eighties spatial access methods for the access on point data have been developed like the grid file (NIEVERGELT et al. 1984). Well known examples of spatial access methods for rectangle data are the R-Tree (GUTTMAN 1984) and the R*-Tree (BECKMANN et al. 1990). One of the most used database queries in today's map-based GIS is the region query which should directly be supported by a spatial access method. A typical example of that type of query is: "Return all cities which are inside the query window and have more than 100.000 inhabitants". Figure 2.8 shows a region query for cities in North Rhine Westphalia, Germany.

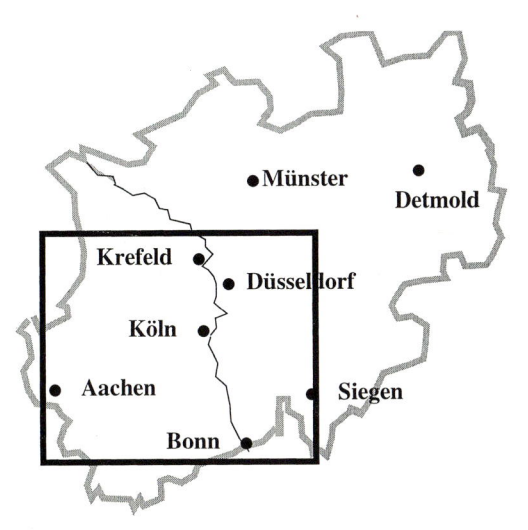

Fig. 2.8 Region query for cities of North Rhine Westphalia, Germany

The temporal access -which mostly is one-dimensional- can be supported with traditional one-dimensional access methods like the B-Tree (BAYER and MC CREIGHT 1972), at least as long as the temporal access is not applied in combination with the spatial access.

2.3.2 Spatial Access Methods

WIDMAYER (1991) as well as NIEVERGELT and WIDMAYER (1997) give a well structured introduction into spatial access methods. In the last reference they describe their "three step model of spatial object retrieval" in detail. The three relevant steps in this model for retrieval of spatial objects are *"cell addressing"*, the *"coarse filter"* and the *"fine filter"*. Cell addressing maps a database query into a set of query cells. The coarse filter consists of a set of query cells that determines the candidate objects in secondary memory. The candidate objects are those objects that are qualified by the query predicate. Objects are qualified, if they are inside the query cells. Technically, the coarse filter is realized with the access on the approximated objects, mostly the minimal circumscribing bounding boxes of the objects. During the execution of the fine filter each object is exactly compared in main memory with the query predicate.

The interface of spatial access methods usually consists of a *retrieve* function which is responsible for the retrieval of object sets. Furthermore, an *insert* function and a *delete* function are provided for adding and removing of single objects. Spatial access methods are often used for the preselection of costly geometric operations like the intersection of geo-objects, the neighbour search or the so called spatial join (BRINKHOFF et al. 1993). The spatial join operation compares two object sets according to a geometric predicate. The result contains the object pairs that meet the query predicate. The preselection is realized with the axis-parallel minimal circumscribing rectangles as a spatial key of the real geometries. An important aspect is to achieve the best possible physical clustering of the objects on secondary memory. This means that the geo-objects have to be stored as closely as possible according to their spatial neighbourhood in 2D or 3D space.

We subdivide spatial access methods into three different groups: the first group are access methods that subdivide space recursively into four regions (2D space) and eight regions (3D space), respectively. These methods are also known as "space-driven partitions". The second group consists of access methods for the management of n-dimensional points. Higher dimensional objects are transformed into points. In the third group of spatial access methods rectangles are managed. Space overlapping rectangles are allowed. These access methods are also called "data-driven partitions". We present an example for each of these groups. Modern spatial access methods typically are refinements of these "traditional" access methods. They distinguish themselves in optimized algorithms, e.g. for the splitting of a node which occurs, if a node is filled up to its maximum number of entries.

2.3.2.1 Quadtree on Top of B^*-Tree

The idea of the quadtrees (FINKEL and BENTLEY 1974; SAMET 1990) is to map more-dimensional geometries to a one-dimensional key. These keys are managed by a one-dimensional access method as they are provided by today's commercial database management systems (ORENSTEIN 1986). The two-dimensional key is stored on disk lexicographically ordered as a one-dimensional key. The quadtree divides the data space regularly into four sub--spaces .The root node has the code "0" and represents the complete data space. The nodes on the first level of the tree represent a quarter of the data space etc..

Chapter 2. Fundamental Principles 25

The problem is to map the data in such a way that more-dimensional region queries are well supported. Thus an optimal preservation of the regions order on external storage addresses has to be achieved. MORTON (1966) has proposed the z-order (Fig. 2.9).

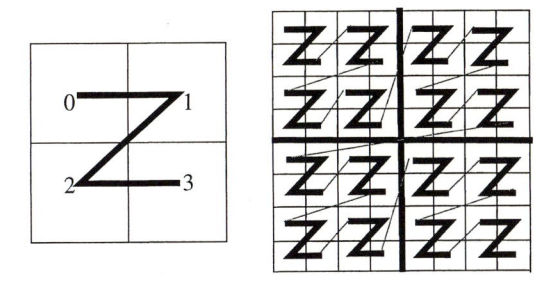

Fig. 2.9 Z-order

The z-order, which determines the quadtree codes, is a mapping from the geometries in the plane to a one-dimensional ordered key. The codification for the nodes of the first tree level is "0" for the northwest, "1" for the northeast, "2" for the southwest and "3" for the southeast data sub-space. Each node of the quadtree including the interior nodes, should correspond to a database page on secondary memory. Such a database page consists of all objects that are completely inside the data spaces of the node, but that are not completely inside one of the data sub-spaces of the next finer level.

In a quadtree not only points but also areas can be handled. In this case a quadtree is a unique representation of an arbitrary quadrant with a starting area of edge length $2^n \times 2^n$. The quadtree then is an approximation of the geometry. Each of the four quadrants is again recursively representable as quadtree. Its codification results from the concatenation of the corresponding 0, 1, 2 or 3 with the quadtree code of the predecessor. With each subdivision of a quadrant into its four sub-quadrants the length of the quadtree code grows by 1. The subdivision of an area into its four quadrants corresponds to the addition of one digit to the quadtree code describing this area. Thus the length of a quadtree code is equal to the number of subdivisions of the starting area. A quadtree code of length m can be represented as follows:

$qt_{[4]} = q_1 q_2 \ldots q_m$
with $q_i \in \{0, 1, 2, 3\}$ for quadtree base 4
for $i = 1 \ldots m$

The quadtree code with base 4 can be converted into a binary quadtree code (00, 01, 10, 11) with a digit-by-digit conversion *div 2* and *mod 2*, respectively. The quadtree code for the decimal (x,y)-coordinate of its left upmost point can be computed by the quadtree code of the square that represents the code. Let m be the length of the quadtree code (in the following example m=2) and let 2^n be the edge length of the starting area (in the example n=3). Then the quadtree code can be computed from the (x,y)-coordinates as follows:

$$x = \sum_{k=1}^{m} (q_k \text{ MOD } 2) * 2^{n-k} \quad , \quad y = \sum_{k=1}^{m} (q_k \text{ DIV } 2) * 2^{n-k}$$

We give an example: in Fig. 2.10 x(A) = 4 and y(A) = 2.

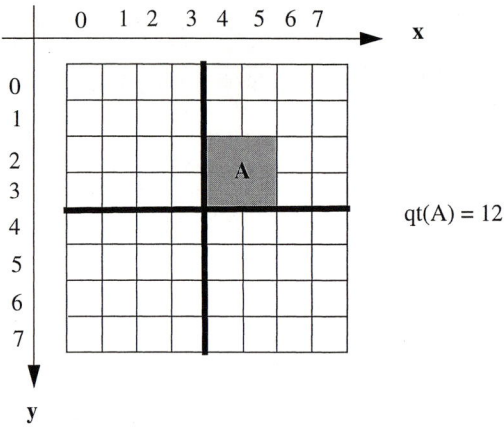

Fig. 2.10 Quadtree code of area A

We outline the algorithm for the so called bit-interleaving:

algorithm *bitInterleaving (x, y)*
{Let x be the x-coordinate and let y be the y-coordinate of the left upmost point of the square that is represented by the quadtree code. Compute the quadtree code qt}

Step 1:   level = 0;
           qt = 0;

Step 2:   while (x > 0) or (y > 0)
            left = x mod 2;
            right = y mod 2;
            newpos = 2 * left + right;
            qt = newpos << (2 * level) + qt   // left shift: variable << # of bit positions
            x = x div 2;
            y = y div 2;
            level = level + 1;

Step 3:   return qt;
end *bitInterleaving*

The quadtree code qt(A) of the example in Fig. 2.10 is computed by the alternate concatenating of the bits from the binary codification ("bit-interleaving") for the x- and y-coordinates:

Chapter 2. Fundamental Principles 27

$$x(A) = 4 = 1 \mid 0 \mid 0$$
$$y(A) = 2 = 0 \mid 1 \mid 0$$
$$qt(A) = \quad 01 \quad 10$$
$$= \quad 1 \quad 2$$

In the shown example the result is a decimal '36' (dual 10 01 00). If we read these digits pair-wise backwards we get 00 01 10. Now we combine each time two dual digits to one decimal digit. Then the result of the quadtree code is qt(A) = 12.

The smallest area that can be represented by a quadtree code is a square with edge length 1. Applied to a base of $2^n \times 2^n$ the corresponding quadtree code has the length n with its maximal resolution. Figure 2.11 shows the tree representation of the quadtree from Fig. 2.10.

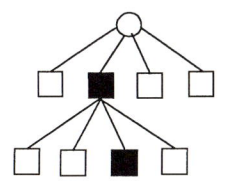

Fig. 2.11 Tree representation of the quadtree from Fig. 2.10

It is particularly convenient if those quadtree codes which are neighbours in the z-order of the lexicographically ordered list are also geometric neighbours. The order is significant for the response time of spatial search queries. The search query can also be converted from the two-dimensional case (query window) into the one-dimensional case. As the number of external storage accesses is $O\sqrt{n}$ for n binary regions, in two-dimensional search queries it is better to map the two-dimensional search region into a set of one-dimensional query regions.

The B^*-Trees (WEDEKIND 1974) are multi searchtrees and a variant of the B-Trees (BAYER and MC CREIGHT 1972). The data sets are explicitly stored in the leave nodes. The root of a B^*-Tree is constructed with a single region from which the next hierarchy level is referenced. Usually a node of the tree corresponds to one page on secondary memory. This means that the number of entries in one node is limited by the page size. All of the leave nodes are on the same level of the tree, i.e. the B*-Tree is balanced.

To realize region queries the quadtree code has to be determined for the query region. The quadtree has to be navigated through level by level. The directory tree (i.e. interior nodes) at most has a logarithmic height. During the insertion and deletion special algorithms ensure that each node is at least filled up to 50%. If this is not the case, nodes have to be melted or split. Such a split can be propagated until the root of the tree is involved.

2.3.2.2 Grid Files

The Grid File (NIEVERGELT et al. 1984) is based upon extensible hashing. A d-dimensional grid is fixed on the d-dimensional universe. Different forms and sizes of the cells are allowed, as the grid does not need to be regular. A grid directory associates one or more cells with data buckets which each time are stored on one page of the secondary memory.

Extended geometries are approximately represented as minimal circumscribing axis-parallel rectangles. The circumscribing rectangles (2-dimensional case) are transformed into points of higher dimension (4-dimensional). For example, each rectangle is represented with its middle coordinates and with the maximal distance to the centre (x_{centre}, y_{centre}, x_{dist}, y_{dist}). A disadvantage of this approach is that after the transformation spatially neighboured objects eventually can be located far away from each other. Figure 2.12 shows an example for the grid directory of a grid file with point data[1].

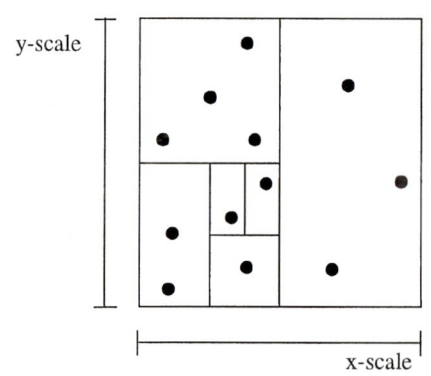

Fig. 2.12 Grid directory of a grid file with points

A two-dimensional region query in the grid file is realized as follows: at first an exact search for one point (e.g. the left upmost) of the query region is executed. Then the element in the grid directory can be computed with the scale indices. The grid itself is managed in main memory as a d-dimensional array (the scales) to guarantee that the data can be found with at most two disk accesses. The first of the two disk accesses is necessary to access the pointer of the searched data, if the data are not yet loaded into main memory. The second access is needed to access the real data.

If a block (node) runs over, it is split. In this case the split dimension (x or y) and the split position (e.g. the centre of the interval) are determined. The new regions should have a square form, if possible. To determine the split dimension the dimension is selected that leads to the longest interval (vertical or horizontal split line). The partition of the grid file is dependent on the insertion order of the data.

---

[1]. More precisely, the points are stored in the data buckets and not in the directory.

2.3.2.3 R-Trees

The R-Tree (GUTTMAN 1984) can be seen as a direct more-dimensional generalization of the B-Tree or the B*-Tree, respectively (BAYER and MC CREIGHT 1972). Each directory region manages a rectangular sub-space of the universe. Furthermore, the rectangles of one tree level can overlap. Each directory region manages a set of regions that are completely inside the rectangle of the directory region. This rectangle is the circumscribing rectangle (bounding box) of the rectangles being located one level below in the tree. Figure 2.13 shows an example of the R-Tree with directory levels.

Each node of the R-Tree should be mapped on one page of the secondary memory to achieve an optimized physical database design. The minimal circumscribing rectangles of the real objects are stored in the leaf nodes. There is a lower and an upper boundary for the number of entries in the nodes of an R-Tree. Let M be the maximal number of entries per node, then $m \leq M/2$ is the lower boundary. The node is deleted and its entries are subdivides to other neighbouring nodes, if the lower boundary falls below this. The root of the R-Tree has at least two entries as long as it is not a leaf node. The height of the R-Tree is balanced in all leaf nodes, i.e. the leaves are automatically kept on the same level during the insertion and deletion operations of nodes. The height of the R-Tree is at most the next higher integer number being less than $(\log_m (N))$ -1 for N nodes and with m for the minimal number of entries of a node. In the worst case capacity for all nodes besides the root is m/M.

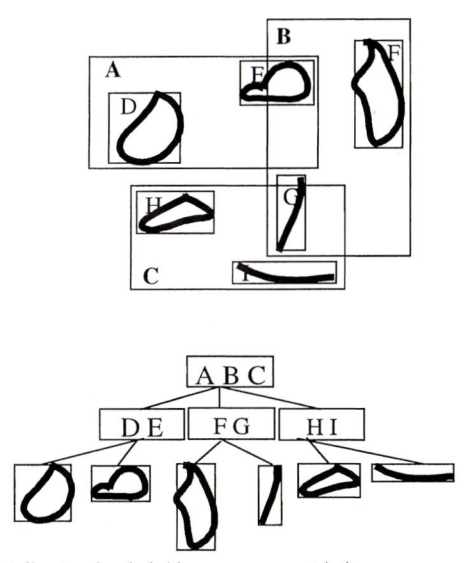

Fig. 2.13 R-Tree with two directory levels (with areas as geometries)

The partition of space and hence the shape of the R-Tree regions is dependent on the insertion order of the rectangles. A node is split, if it runs over. This is the case if the maximal number of entries is exceeded. The new rectangle is inserted into the node whose area of the bounding box has to be enlarged least. The search algorithm of the R-Tree works similar to the search in the B-Tree. Starting from the root of the tree the entries of the nodes are compared with the d-dimensional search interval. The query result consists of the set of candidates that meet the query predicate. The real geometries then have to be tested for exact intersection with the search interval. During the execution of a region query, the more rectangles there are overlapping the more paths there are of the R-Tree that have to be searched through. This multi search has a negative effect on the efficiency of the data retrieval. There are some proposals in literature to optimize the R-Tree like the R^+-Tree (SELLIS et al. 1987) which completely avoids the overlapping of directory regions. The R*-Tree (BECKMANN et al. 1990) computes new regions differently which are created during the insertion of a new entry into an overrunning node. The algorithm does not only consider the size of the regions but also the sum of the perimeters and the size of the overlapping areas. The packed R-Tree (ROUSSOPOULOS and LEIFKER 1985) computes an optimal partition of the universe and a corresponding minimal R-Tree. However, the data has to be known a priori to use packed R--Trees.

The search algorithm is also used to insert objects into the R-Tree. Then the minimal circumscribing rectangle and the object reference are inserted into the right place of the R-Tree. Finally the object is inserted into the leaf node. Eventually the node has to be split. During the deletion of an entry eventually the other entries of the node have to be copied into a temporary node and re-inserted into the tree, if the node is too empty.

Eventually, the insert and delete operations result in an update of the circumscribing rectangles of other nodes up to the root of the tree. A detailed description of the retrieval, deletion and insertion of objects into the R-Tree has been presented in GUTTMAN (1984).

2.3.3 From the Relational to the Object-Oriented Database Design

The Relational Data Model is well suited for the modelling of thematic data. However, there are disadvantages concerning the modelling of geometric and topological data. This has lead to a separated management of these different kinds of data. In today's GIS thematic data are managed relationally in a database management system on secondary memory and the access on spatial, i.e. geometric and topological data is realized with the help of specialized geometric representations in main memory.

Besides the conceptual simplicity of their Relational Data Model (CODD 1970) relational database management systems (RDBMS) have further advantages that explain the general acceptance for many geo-applications:

1. There are theoretically sound approaches for the realization of the query language in a RDBMS;
2. The optimization of queries can be executed automatically (algebraic optimization);
3. Integrity constraints and queries can easily be formulated with a descriptive relational query language.

However, the following disadvantages speak against the use of relational database management systems for GIS:

1. They have no construction for the definition of complex geo-objects (lists, arrays, bags);
2. The data model does not contain enough semantics (e.g. the object behaviour cannot be described);
3. They lack concepts for long transactions;
4. The "impedance mismatch", i.e. set-oriented query results have to be processed tuple-oriented, i.e. data set by data set in the programming language;
5. Relationships between geo-objects can only be constructed by costly "join" operations;
6. They lack extensibility for user-defined data types like geometric data types with suitable operations and multi dimensional access methods.

Especially the last two points lead to efficiency problems that usually cannot be compensated by the advantages of relational databases. That is why extensible database management systems (EDBMS) have been developed since the beginning of the eighties.

What really is required for GIS is the direct storage of hierarchically and network-like structured objects in the database. This is the first step on the way to object-oriented data models. Hierarchically structured objects have already been realized in the NF^2 data model (JAESCHKE and SCHEK 1981; SCHEK and SCHOLL 1986).

Extensible DBMS (EDBMS) originated from the requirement to develop a suitable database support for so called non-standard database applications like GIS, CAD etc.. The idea was to allow the user the extension of standard data types like *integer, real, char* etc. and access methods like the B-Tree at a low system level in the DBMS. Such extensions can especially be used for geometric data types and spatial access methods. The extended data types and access methods should be handled in the same way as the standard types and methods of the DBMS. Extensible DBMS are an approach to "drill" standard DBMS to obtain object-oriented modelling in an efficient way; i.e. the physical management of complex objects should be realized efficiently. EDBMS can be seen as a "bottom-up approach" coming from the database side to obtain object-oriented modelling techniques. Such EDBMS typically consist of an object storage manager for the physical management of complex objects at the deepest system level. On top of the object storage manager the data model is realized within an object manager. DITTRICH (1986) called such systems "structural object-oriented", because the behaviour of the objects cannot be described in their data model.

Object-oriented data modelling and object-oriented databases have some further advantages that makes them particularly attractive for the modelling and management of geoscientific data:

1. The ability to model spatial and non-spatial data within a unified data model;
2. 1:1 mapping between the database types and user-defined data types;
3. Representation of the objects' behaviour;
4. Inheritance;
5. Polymorphism.

As we have already argued in chapter 2.1.2.2, the integrated modelling of thematic data and geometric or topological data in the object-oriented paradigm can be realized in two ways: the first is to model geometry as a data member of a thematic geo-object class *(part-of relationship)*. Alternatively, a thematic class can be defined as a specialization of a spatial geo-object class. This means that each class of a specific geo-object class inherits the spatial functionality of the general geo-object class. It adds its own specific data members and methods. Furthermore, in the object-oriented paradigm collections can be defined to realize an efficient spatial access for geo-objects that are spatially related. In object-oriented databases the user-defined classes are directly translated into database classes. Therefore user-defined and database classes fit well together and there is no "impedance mismatch" between both representations as is the case with relational databases. Furthermore, in the database classes the structure and the behaviour of the objects are modelled together as a unit and the concept of inheritance allows the geoscientist the direct extension of already existing classes. Finally, the concept of polymorphism enables the implementation of different function implementations, e.g. to realize different implementations for the intersection of geometric objects like circles, rectangles or triangles. Polymorphism reduces the size of the code in GIS applications significantly.

Object-oriented DBMS (OODBMS) have been developed by a "top-down approach" with the idea to enrich object-oriented programming languages with persistent objects that should "survive" the duration of one programme execution. The idea was on the one hand to use proved database technology and on the other hand to benefit from object-oriented data modelling. In the data model of an OODBMS persistent and non-persistent (transient) objects are equally treated, i.e. no costly translations are necessary between them. We speak of a "transparent view" for the user concerning persistent and transient objects. The data model -e.g. realized with the C++ programming language- which is equivalent to the internal database model avoids the "impedance mismatch" between programming language and query language. Versions and configurations of objects in a geo-process can at least be managed by some of today's OODBMS. Seen from a GIS perspective, the most obvious disadvantages of the OODBMS realized with the "top-down approach" like GemStone, VERSANT, Poet, ONTOS, ObjectStore etc. are the following: the lack of suitable spatial access methods at deep system levels[1], the mostly lacking or only partially realized declarative ad-hoc query language and the lack of a suitable query optimization. Probably the boundaries between the systems which are realized with the bottom-up and the top-down approach will become blurred in future OODBMS as is already the case with O_2 (DEUX 1990). It is likely that both approaches will be combined in one system to use the advantages of both approaches.

2.4 Geodata Analysis

2.4.1 Elementary Geometric Algorithms for GIS Applications

The management and processing of geometries is a central task for GIS. Spatial queries like the well known window query (e.g. "return all cities inside a query window"), the point query (e.g. "return all maps that contain Bonn") or the intersecion query (e.g. "return all motorways

---

[1]. A first approach concerning the extension of spatial access methods realized in a commercial OODBMS is the spatial object manager of ObjectStore.

in North Rhine Westphalia that cross rivers") need geometric algorithms for their response. Approximated geometries like minimally circumscribing rectangles are used in a first step as a coarse filter to reduce the search space. In a second refinement step the real geometries are compared with the query predicate.

We can subdivide geometric problems in GIS applications into three types:

1. One-object problems;
2. Object set problems;
3. Search problems.

In one-object problems single objects are processed. An example is the triangulation of a polygon. In object set problems a property of an object set is determined. Examples are the convex hull of a point set or the determination of all pairs of line segments that are intersecting (so called segment intersection problem). Finally a search problem is a problem in which the relationship of objects in an object set to a query object is determined. For search problems we need data structures that enable efficient search. Examples of search problems are the determination of all rectangles from a set of rectangles that contain a point p (point query) or of all rectangles that intersect a query rectangle r (intersect query).

General segment intersection problem

Following the ideas of GÜTING (1992) for right-angled geometries we introduce the solution of the *general segment intesection problem with a plane-sweep* and a *divide-and-conquer* algorithm. The goal of these two techniques developed in "computational geometry" (SHAMOS 1978) is to improve the runtime of geometric algorithms and to handle geometric problems simpler for concrete practical problems.

With the *plane-sweep technique* we "sweep" with a line over the plane from the left to the right side. The goal is to handle only a limited number of ordered objects at every time step of the algorithm. The "processed objects" do not have to be handled more than one time. Plane-sweep reduces a k-dimensional set problem to a (k-1)-dimensional search problem.

With the *divide-and-conquer-technique* an object set is subdivided to apply the geometric problem recursively to subsets of the object set. The partition is determined by the geometric location of the objects in space. Divide-and-conquer reduces a k-dimensional set problem to a (k-1)-dimensional set problem.

Consider a set of arbitrarily oriented segments in the plane. We search all pairs of intersecting segments. For simplicity we assume two essential restrictions (Fig. 2.14):

1. Maximally two segments intersect in the same point;
2. The set does not consist vertical segments.

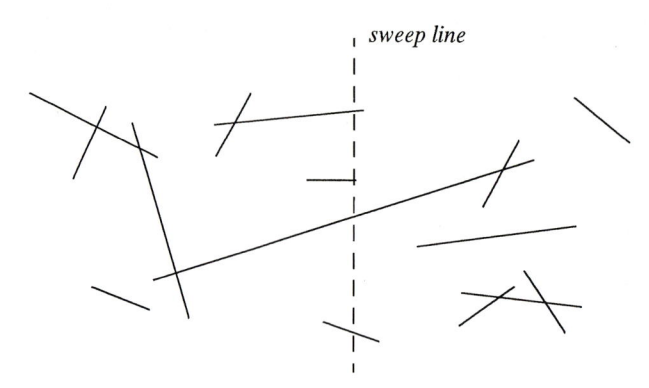

Fig. 2.14 Plane-sweep solution

Solution with plane-sweep:

We make the following observations:

1. To every time point the *active* segments are the set of segments that intersect the sweep line;
2. The y-order of the intersection points of the active segments with the sweep line define an *order;*

the sequence of the active segments ordered by the y-coordinates only changes during the sequence, if:

1. The left end point of a segment is reached;
2. The right end point of a segment is reached;
3. The intersection point of two segments is reached.

Points 1) - 3) are the *stop points* or *stations* of the plane-sweep.

Two segments can only intersect, if they have been neighbours before in the sequence of active segments.

Algorithm:

1. Construct the queue Q of all stop points that have to be processed.
   Q = the set of all end points of segments ordered by the x-coordinates.
   S = sequence of all active segments, $S := \{\}$.

2. Take the first element q from Q, respectively:

   a) q is the left end point of a segment t (Fig. 2.15):
   $\Rightarrow$ insert t according to the y-coordinate into S
   $\Rightarrow$ let s, u be the segments that are directly above and below t, respectively:

does t intersect s or u?
⇒ insert each intersection point into Q ordered by its x-coordinate.

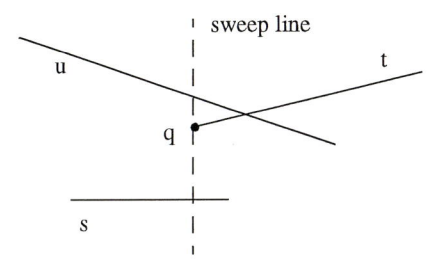

Fig. 2.15 Processing of the left end point of a segment

before	after
$Q = <t_l, s_r, u_r>$	$Q = <s_r, (t, u), u_r>$
$S = <s, u>$	$S = <s, t, u>$

b) q is the right end point of a segment t (Fig. 2.16):
⇒ delete t from S
⇒ do the segments s and u that have been above and below t, intersect each other?
⇒ insert each intersection point into Q, ordered by its x-coordinate.

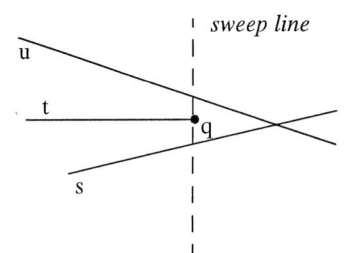

Fig. 2.16 Processing of the right end point of a segment

before	after
$Q = <t_l, u_r>$	$Q = <(s, u), u_r>$
$S = <s, t, u>$	$S = <s, u>$

c) q is intersection point of the two segments t and t' (Fig. 2.17):
   $\Rightarrow$ return the pair (t, t')
   $\Rightarrow$ change t and t' in S
   $\Rightarrow$ let s, u be the neighboured segments above and below t, t', respectively:
   does t intersect s or does t' intersect u?
   $\Rightarrow$ insert each intersection point into Q, ordered by its x-coordinate.

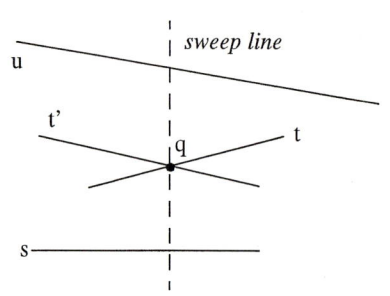

Fig. 2.17 Processing of the intersection point of two segments

before	after
$Q = <(t, t'), s_r, t'_r, u_r, t_r>$	$Q = <s_r, t'_r, u_r, t_r>$
$S = <s, t, t', u>$	$S = <s, t', t, u>$

For plane-sweep algorithms we need two data structures:

1. A *sweep line status structure* S that can be realized as a balanced tree. Each node contains the coefficients of the line equation of the segment.
2. A queue Q, that provides the following operations:

   (i) Deletion of the element with the smallest x-coordinate;
   (ii) Insertion of the element with arbitrary x-coordinate.

One of the data types which is well suited for this task is the *priority queue* (see also GÜTING 1992).

Time complexity:

Data structure S as well as Q can be realized as a tree. Thus each operation on S or Q costs $O(\log n)$ time with n segments. The sweep at most has $n+k$ stop points with k intersection points, i.e. the complete costs are $O[(n+k) \log n]$ time.

Space complexity:

The sweep at most has $n+k$ stop points, i.e. the costs for Q are $O(n+k)$.

Chapter 2. Fundamental Principles

Solution with divide-and-conquer:

In the following let S be the set of arbitrarily oriented segments with the restrictions as before.

Divide: choose the x-coordinate x_m that subdivides S into two roughly equally sized sub-sets S_1 and S_2

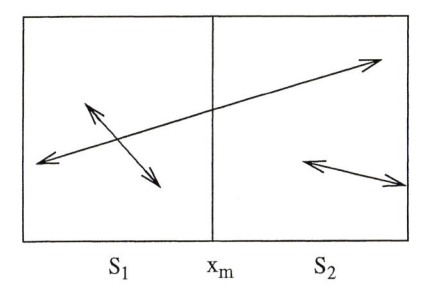

Fig. 2.18 Divide-and-conquer solution

Conquer: intersection(S_1), intersection(S_2)

The *recursive invariant* for intersection(S_i) runs: intersection(S_i) returns all intersections between the segments that are represented in S_i.

Merge: during the *merge*-step only the intersections between the segments of S_1 and the segments of S_2 have to be computed.

We distinguish the following cases for a single segment h of S_1:

Case a)   Both end points of h are inside S_1 (Fig. 2.19).

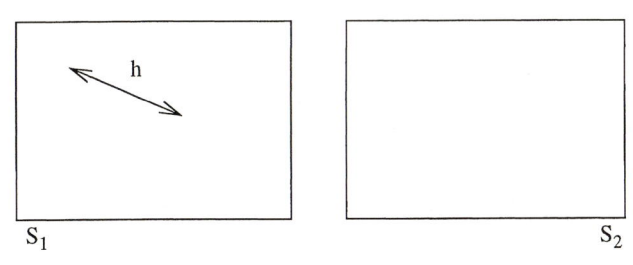

Fig. 2.19 Case a): both end points of h are inside S_1

h does not intersect any segment of S_2.

Case **b)**   Only the right end point is inside S_1 (Fig. 2.20).

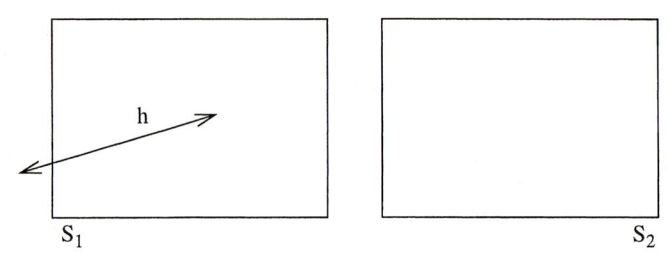

Fig. 2.20 Case b): only the right end point is inside S_1

h does not intersect any segment of S_2.

Case **c)**   Only the left end point is inside S_1.
            c-1)   The right end point is inside S_2.

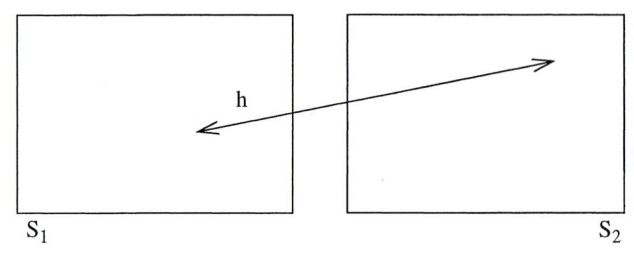

Fig. 2.21 Case c-1): the right endpoint is inside S_2

True to the recursive invariant: all intersections between h and the segments in S_2 are already computed.

c-2) The right end point is right of S_2.

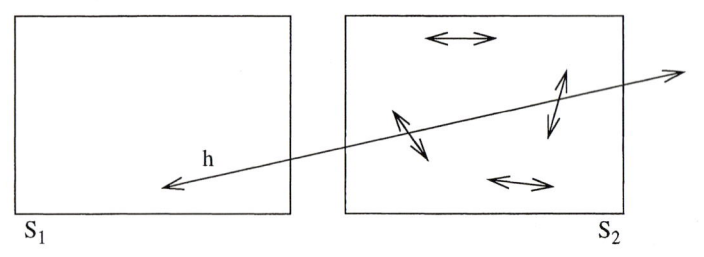

Fig. 2.22 Case c-2): the right end point is right of S_2

h totally "crosses" S_2. h potentially intersects a certain sub-set of segments in S_2. The merge-step determines this set and the intersecting pairs.

Merge: - determine all left end points of S_1 whose partners (right end point) are neither inside S_1 nor inside S_2.
- determine the subset of S_2 that potentially is intersecting and find the intersection pairs.
- execute both steps analogously for the right end points of S_2.

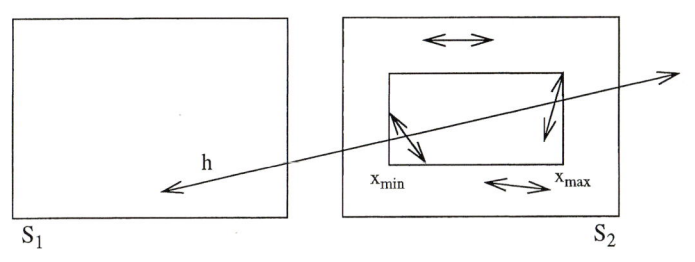

Fig. 2.23 Merge-step

Let M be the set of segments.

$$S = \{ (x_1, (x_1, y_1, x_2 - x_1, y_2 - y_1)) \mid (x_1, y_1, x_2, y_2) \in M \} \cup \\ \{ (x_2, (x_2, y_2, x_1 - x_2, y_1 - y_2)) \mid (x_1, y_1, x_2, y_2) \in M \}$$

The set of intersecting segments is completely computed by the merge step. To reduce its time complexity which is $O(n^2)$, the segments should be stored in a well suited spatial data structure like the R*-Baum (BECKMANN et al. 1990).

2.4.2 Visualization of Spatial and Temporal Data

In the context of GIS spatial and particularly geometric data are often denoted as "graphical data". To avoid this confusing terminology we distinguish, seen from a database point of view, between *graphical* and *geometric data* in the following way: geometric data describe the geometry and especially the points and the coordinates of spatial objects. Graphical data meet the visualization of the objects, e.g. in a map. For example, they describe the graphical representation of a city (as a point or an area), the thickness of a motorway line or the pattern or colour of industrial regions in a map.

During the last years the visualization of spatial data has been recognized as an important instrument of GIS for the clear presentation of large data sets. Essential features can be extracted and complex geometries and their spatial relationships can be made more accessible (HEARNSHAW and UNWIN 1994). With the *visualization in scientific computing (ViSC)* a new research field has been established. Its objectives are to examine problems that deal with data visualization and visual interactive data manipulation especially in 3D space up to the animation of objects ("4D") (MC CORMIC et al. 1987). It has been recognized that it is important to make geodata accessible in a fast and clear way during all steps of their processing. Therefore an *interactive graphics environment* is necessary. In 3D applications like geology spatial objects have to be manipulated interactively. Thus it is necessary to embed a 3D/4D visualization component into the graphical user interface. We can distinguish between two techniques for the visualization of objects: on the one hand we observe the visualization of the geoscientific objects like geological layers and faults with shading methods. On the other hand there is the visualization of the spatial representation like triangle networks represented with a wire-edge representation. This technique for visualization represents the internal topological structure of the data. Both techniques are important for GIS. The first one pregnantly represents the spatial situation and provides a good overview, whereas the second can efficiently be used for a detailed structure analysis.

The formulation of spatial and temporal database queries should directly be supported by the visualization software. Examples are the visualization of geometric 3D operations like the intersection of surfaces in 3D space with a query box or the intersection of arbitrary geological boundary surfaces or fault surfaces in space and time. The "mouse" can directly be used for the selection of the single objects. In more complex database queries like "return the volume of all geological bodies that are at least 300 m under the surface of the Earth and that existed in the Tertiary" the visualization is only suited to present the general spatial situation of the objects. The formulation of the database query, however, should be done with a database query language.

In the field of 2D GIS query languages have been developed that particularly support the graphical query formulation and the output of database queries. Examples are Mapquery (FRANK 1982), Pictoral SQL (ROUSSOPOULOS and LEIFKER 1985), Geo-SAL (SVENSSON and ZHEXUE 1991), Spatial SQL (EGENHOFER 1994). In 3D space and 4D, however, as far as the author is aware, no such approaches have been applied.

Of course the result of spatial database queries should again be visualized. The geoscientific expert can use the query result later for spatial integrity checks of her/his initial model. This means that the visualization also serves as a quality check for geoscientific models. For the intersection algorithms mentioned above it is not possible to evaluate the result efficiently in a non-visual way. The geometric 3D operations should algebraically be closed so that the data type of the visualized result can again be used as the input data type of new 3D operations.

Geoinformation systems are increasingly used as visualization tools. This is a consequence of the fast development in hardware and visualization software during the last years. However, the connection with database management systems was not well examined until the present day. The problem has been the efficient coupling between graphics and database software. We will pick up this point again in chapter 5.2.6.

Chapter 3

Examples of Today's Geoinformation Systems

In this chapter the state of the art of some commercial and prototypical geoinformation systems is presented. Finally the deficiencies are shown by comparing the systems.

3.1 Commercial Systems

In the following we present representatives of commercial GIS and of such GIS that have been developed at universities. The selection of the GIS has been done subjectively. It is not to be seen as positive or negative valuation for or against GIS that have or have not been cited in the list. We also do not claim any completeness of the descriptions.

3.1.1 ARC/INFO

ARC/INFO originated from a city map system that had been developed by ESRI in 1969 for the city of Los Angeles. The system is widely spread and available on many hardware and operation systems platforms. The functionality of ARC/INFO has grown over the years with the requirements of its users. Today it is a very extensive GIS and many application shells have been developed on top of ARC/INFO. The kernel system covers more than 2500 commands. Of course there are also "historically grown modules" in ARC/INFO that partially cover the functionality of later developed modules. However, obviously this is the case with any software that is not brand new. The geometric and topological data are processed with ARC. INFO or a coupled relational database management system is responsible for the management of the thematic data. ARC/INFO is a "layer-based" GIS, i.e. different maps can easily be intersected. Examples are the intersection of "coverages", "GRIDS" or "lattices" which produces a new map, respectively.

User Interface:

Concerning the user interface the user can select between three ways of data input with ARC/INFO:

1. In a command line;
2. With the ARC Macro Language (AML);
3. With the user interface ARCTOOLS, a graphical user interface (GUI) which for the most part is programmed in AML and based upon standards of graphical user interfaces like OPEN LOOK, OSF/MOTIF and Microsoft Windows NT.

In the following we give a short typical ARC/INFO session with AML (Fig. 3.1).

```
DISPLAY 9999 1                              // open window on the complete screen
POLYGRID soiluse soilgrid area              // convert polygons into a grid
GRID                                        // call the GRID module
depositarea = zonalarea ( soilgrid, deposit, DATA)
                                            // find the max. deposit in each region
if ( soilgrid < 200 & depositarea > 400 ) outgrid = 1      // intersection
KILL depositarea                            // delete the intermediate results
KILL soilgrid
MAPExtent outgrid                           // visualization
GRIDShade outgrid
QUIT
```

Fig. 3.1 Example of an ARC/INFO session

Let us assume that a grid should be generated from the poly(gon) coverage *"soiluse"* and the grid *"deposit"* that shows small regions with the soil use and with high deposit. The commands can be read from a file or they can be called by AML with the "arc-prompt".

Visualization:

For the interactive visualization of maps, images and spatial data the tool Arc/View can be used. It also allows to use raster pictures in different formats as background for vector maps. Furthermore, query results like the selection of thematic attributes of polygons can be viewed graphically. An own programming language provides the programming of Arc/View functions and the construction of Macros. The functionality of Arc/View is extensive; it covers from integrated table calculation to professional graphics. In the meanwhile Arc/View can be seen as an own GIS that can compute complex analysis.

Data Input:

The user has two possibilities for the input of geometric data:

1. Digitalization;
2. Reading external data formats.

Chapter 3. Examples of Today's Geoinformation Systems 43

1) Digitalization

The data input in ARC/INFO takes place on the digitalization tablet with the ARCEDIT editor. To digitize the data the user first specifies a map with so called *tics,* i.e. the reference points of a coverage. Then the digitalization process can start. After that the correction of digitalization errors takes place (overshoots, undershoots, dangle nodes). ARC/INFO considers a certain error tolerance with the so called "snap-distance" which applies an epsilon environment for digitalized points. This distance can be put at a certain value by default.

2) Reading External Data Formats

With ARC/INFO all typical GIS and CAD exchange formats and pictures (bitmaps) can be read.

Data Modelling:

The data modelling of ARC/INFO is based on the so called georelational data model (MOREHOUSE 1985) which describes thematic, geometric and topological data with tables. These data are acquired as map coverages. It is possible to reference several coverages to one workspace in main memory. As already mentioned, INFO is responsible for the management of thematic data. The management of geometric and topological data takes place in ARC.

Like in other layer-based GIS *complex geo-objects* can be defined in ARC/INFO with the overlay of maps like coverages or grids. The thematic data of these objects, however, have to be composed by relational join operators. Furthermore, only one attribute can be defined per coverage.

Data Management:

In ARC, each coverage has pointers to some tables for the management of the geometry and topology (so called feature attribute tables). The topology of a polygon coverage is managed in the following way. For each arc its left and right polygon and its start and end point are referenced and each polygon manages a list of its arcs. Furthermore, a user-defined set of geographical features can be composed to a theme. The INFO tables can alternatively be handled with a relational database management system. This means that also interactive database queries formulated with SQL (Structured Query Language) are possible.

ARC/INFO knows predefined attributes for polygons like AREA, PERIMETER, Coverage--ID etc.. They are automatically generated by ARC/INFO during the generation of a coverage.

In ARC the geometric base elements (features) are managed with *polygons* (non-overlapping partition of space), *regions* (set of eventually overlapping polygons), *routes* (lines), *arcs* (line segments) and different variants of points like *tics* (reference points of a coverage), *nodes* (points which are fixed during the digitalization process), *label points* (fixed points in a map that serve as lettering, see also Fig. 3.2 and 3.3).

Arc   *Route*   *Polygon*   *Regions*

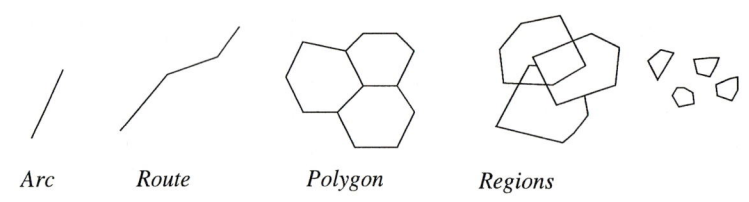

Fig. 3.2 Line and area-like primitives

Furthermore, ARC/INFO provides the feature type *annotation* that enables the lettering of text in map elements. The further processing of spatial data can be executed with further ARC/INFO modules like ARCEDIT, ARCPLOT, GRID and TIN.

City of Bonn

Label Point   *Nodes*   *Tics*

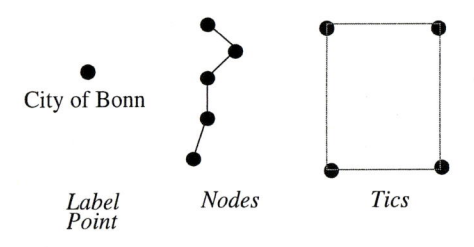

Fig. 3.3 Point-like primitives

A disadvantage of all closed GIS architectures is that the coordinates, e.g. of polygons in a map, are stored with an internal data format. Thus it is difficult to get the geometric information of these data. However, the coordinate values can be edited by moving them on the screen.

The combination of more than one coverage can be executed with the intersection function. The user can select -depending on the type of maps that are to be intersected like polygon/ point, polygon/line or polygon/polygon- between the different functions *union, intersection* and *identity*. During the intersection the referenced thematic attribute of the first map sheet is unified with that of the second map sheet. An example for a result of the intersection is the two attribute values "soil use is forest" and "soil encumbrance is small". Therefore also queries on attributes of different coverages are possible. During the combination of map sheets a new map sheet (coverage) is temporarily generated in main memory.

Raster and Vector Processing:

ARC/INFO mainly covers the vector data processing, however there are also functions for the conversion of vector and raster data and vice versa. The vectorization of a specified raster means to combine raster cells with the same attribute value to a polygon. Areas that are classified in this way as polygons can also have thematic attributes. In the opposite way, the rasterization, the single polygons are decomposed into raster cells with the same attribute value. The corresponding ARC/INFO functions are called *gridpoly* and *polygrid,* respectively (see Fig. 3.4).

1	1	1	1	2	2	2	2
1	1	1	1	2	2	2	2
1	1	3	3	2	2	2	2
3	3	3	3	3	3	2	2

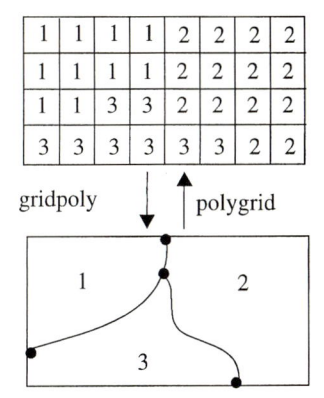

gridpoly ↓ ↑ polygrid

Fig. 3.4 Vectorization and Rasterization

Spatial Queries and Analysis:

To the knowledge of the author, hitherto ARC/INFO does not provide a spatial access method for the efficient support of spatial region queries. However, there are some spatial operations like *clip* for the clipping of regions in a map sheet or *split* for the partitioning of a map sheet in a number of smaller map sheets; *mapjoin* effects the merging of adjacent polygons with reconstruction of the topology, i.e. the neighbourhoods of the polygons. The *buffer*-function generates regions around points, lines or polygons in point, line or polygonal map sheets. Such buffers are very useful in many applications like the detection of risk zones in environmental monitoring.

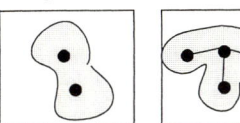

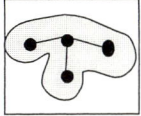

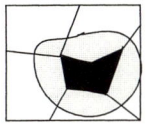

Fig. 3.5 Buffer regions around points, lines and polygons

Other essential topological and metric functions are the determination of the next geometry to a given point *(near)* and the determination of the distance between two points *(pointdist)*. The *thiessen* function is very useful for the generation of voronoi diagrams (thiessen polygons) from a point set. The conversion of a GRID into a polygonal map sheet is also often used to process grid data in the GIS.

3D/4D Processing:

The functionality for 3D/4D processing is small in today's available GIS. ARC/INFO provides an automatic triangulation of a point set on a surface for the generation of digital terrain models. The triangulation can be driven by constraints, the so called soft and hard breaklines, like predetermined breaking edges in geology. In the TIN representation each triangle contains three nodes and three pointers to the adjacent triangles. The disadvantage of the TIN representation for geological applications is that multiple z-values are not allowed. This is no problem for the representation of digital terrain models, but in geology also complex forms like salt domes have to be modelled (Fig. 3.6).

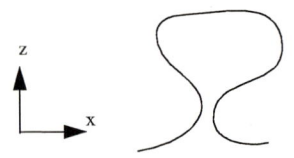

Fig. 3.6 Multiple z-values for the modelling of a salt dome in geology (sketch)

The 3D spatial database queries or intersections of surfaces with volumes in 3D space are not provided by ARC/INFO. For the management of temporal objects an attribute "time" could be added to the thematic attributes of the data. Different scenes could be represented in different layers of a map. However, the time management cannot flexibly be arranged for different objects and the management of the scene would have to be handled by the user.

3.1.2 SYSTEM 9

SYSTEM 9 (VAN ECK and UFFER 1989; SYSTEM 9 1992) was one of the first GIS that did not handle different layers of a map, but objects (features) with thematic and geometric or topological attributes in a relational database management system.

User interface:

Besides using the menue driven user interface in SYSTEM 9 one can create macros and menues with the so called ACL scripts. It is possible to create commands from a sequence of commands with the ACL (Application Control Language) and to call GIS functions. For database queries SYSTEM 9 offers an SQL interface. The database queries can also contain spatial predicates like *contains* or *intersect*. Furthermore, the complete functionality of today's RDBMS are provided and DB-reports simplify the presentation of the database results.

Visualization:

Spatial objects in SYSTEM 9 can be visualized and interactively processed with the "graphics handler" that contains a graphics editor and a graphics library with predefined graphics styles. To keep the graphical representation of a feature independent of the physical database representation the graphical representation is not directly stored in the database, but only the pointer of a feature class to the corresponding graphics style. The system can also easily be extended by new graphics styles.

Data Input:

The data input with SYSTEM 9 can be carried out by digitalization or with special data input and editing software for topological and thematic data. Data exchange with most GIS and CAD data formats and with relational DBMS is provided. Furthermore, 3D data can be captured by a photogrammetric workstation.

Data Modelling:

Based on the relational data model complex geometric object classes can be modelled in SYSTEM 9 like an object class for triangulated irregular networks (TIN). The objects consist of the geometric basis elements ("primitives") *node* (point), *line* (segment, arc, circle, spline) and *surface* (regions inclusively "islands"). Additionally, in SYSTEM 9 the primitive *spaghetti* exists for lines without topological relationships to their neighbouring lines or areas. The primitives can be used by more than one object (so called shared primitives).

Data Management:

Unlike the layer-based systems in SYSTEM 9 spatial data as well as thematic data are managed as "features" in an RDBMS with their spatial relationships. Geometry is treated as an attribute of the objects. That is why in SYSTEM 9 no coordination between the different data management structures of spatial and thematic data is necessary. An arbitrary number of attributes can be attached to any object ("feature"). Finally a sheetless management of the data is provided.

Spatial Queries and Analysis:

As we have already mentioned spatial database queries with SYSTEM 9 can be formulated with an SQL interface or with a 4GL. Selections and projections are provided as well as queries analogous to the relational join.

A toolbox-based approach supports flexible and extensible modules that can process sets of objects for spatial analysis. The user can compose his/her own specific solutions with these modules. The database queries are realized independently of the graphical representation of the objects; i.e. they are not - as is the case with other GIS- dependent on fixed layers. This concept enables a flexible logical combination (AND, OR, NOT) of whole map sheets as well as of single objects with different geometries (point, line, area).

A peculiarity of SYSTEM 9 is that a photogrammetric evaluation system is integrated into the GIS. Furthermore, a tool for network analysis, e.g. for roads, supports algorithms for the search of the shortest path between two locations (graph algorithms) and other GIS algorithms.

3D/4D Processing:

In SYSTEM 9 surfaces in 3D space can be visualized in a wire frame or in a shading representation. The management of temporal data could be realized in the same way as with ARC/INFO, i.e. with an additional time attribute. The version control, however, then would have to be handled by the user him-/herself.

3.1.3 SMALLWORLD GIS

SMALLWORLD GIS can be considered as a toolbox for the construction of GIS applications. The idea is that the user can extend or modify as many system components as possible. Unfortunately, this approach is not applicable to the system kernel including the RDBMS. From a commercial point of view, this is of course understandable.

There have not been developed yet so many application shells for SMALLWORLD GIS as for ARC/INFO. However, there are specific shells for cadaster, electricity, gas, water, district heating, industrial effluents and others. Traditionally the strong side of SMALLWORLD GIS is the management and processing of line data like electricity lines[1].

One of the advantages of SMALLWORLD GIS is the transparent coupling with external RDBMS realized by the "virtual database interface" for different RDBMS. SMALLWORLD GIS has a client/server architecture. This system architecture guarantees the portability to heterogeneous system environments. However, the independency of the system platform has to be paid with a lower performance.

User Interface:

SMALLWORLD GIS provides an environment for system developers and its own object--oriented programming language ("Magic") which is similar to Smalltak. In this respect SMALLWORLD GIS is a step further than traditional GIS which often only provide a macro language. For some years the "Glazier" tool has been available which supports the interactive generation of menues. Furthermore, tools like a class and methods browser, a CASE tool and presentation graphics have been developed. SMALLWORLD view supports the browsing of objects in SMALLWORLD databases.

Data Modelling:

In SMALLWORLD GIS thematic, geometric and topological data are united in objects. At the topological level nodes, edges and polygons can be defined. At the geometric level points, chains and areas are the essential primitives. A chain consists of a set of connected chains and an area can potentially consist of a set of non-connecting polygons.

Data Input:

In SMALLWORLD GIS the digitalization of vector data is directly executed at the screen and the semi automatic vectorization of scanned raster maps is preferred because of the more simple error correcting tools. Scanned raster maps can be converted with a semi automatic vectorization into vector data. One of the advantages of the system is that sheetless digitalization is provided. Thus objects crossing different sheets do not have to be handled separately.

Data Management:

In contrary to traditional layer-based GIS in SMALLWORLD GIS a hierarchical class library of objects is supported which includes thematic, geometric and topological data with corresponding methods. The SMALLWORLD database is based on the relational data model with object-oriented extensions for the management of spatial data.

A special feature of SMALLWORLD GIS is the support of long transactions with a version management. Traditional GIS only support short transactions which are based on locking

---

[1]. SMALLWORLD Systems GmbH was founded in 1990 as a cooperation between SMALLWORLD Systems Ltd. (Cambridge, England) and the Gelsenwasser AG (Gelsenkirchen, Germany).

mechanisms of available RDBMS. Another advantage of the SMALLWORLD solution is that transactions for spatial data are supported as well as transactions for non-spatial, i.e. for thematic data. Variants of the database or sub-sets of it can be copied as private and hierarchically structured versions. They are managed by a versioned B-Tree. Within a variant, short time transactions can be defined. The user is responsible for the consistency of the data and for the combination of the different versions before the check in; i.e. the users have to decide which version should be the major version for further processing.

Raster and Vector Processing:

Like most of the common GIS SMALLWORLD GIS also is a hybrid GIS. It provides functions for the combination of raster and vector data. The processing for both data types is supported.

Spatial Queries and Analysis:

In SMALLWORLD GIS spatial 2D region queries are supported by a quadtree which is used as a spatial access method. Small overlaps of the quadrants are allowed, i.e. there are no hard requirements to decide in which quadrant the geometries should be stored. SMALLWORLD GIS enables the spatial clustering of the objects on disk. However, the dynamic management of objects is not supported by the quadtree. The spatial neighbourhood of the objects in the quadtree is only considered at the time of the generation of the quadtree. The spatial key is stored in the identity (Id) of the object and cannot be changed afterwards. This pragmatic approach, however, is acceptable for many geo-applications, because updates are performed scarcely.

3D/4D Processing:

In SMALLWORLD GIS a module exists for the processing of raster data like digital terrain models (DEM) and for the generation of triangle meshes from point sets. The DEM also can be visualized as 2.5D models. As in the other GIS there is no advanced 3D functionality realized so far. Temporal objects can be managed by a version management as we have described before.

3.2 Research Prototypes

3.2.1 GEO^{++}

GEO^{++} (VIJLBRIEF and OOSTEROM 1992) is a GIS frontend with interactive user-interface for the object-relational DBMS Postgres (STONEBRAKER and KEMNITZ 1991). Postgres already had an R-Tree (GUTTMAN 1984) integrated as spatial database class. Geometric data types like *point, lseg, path* (for polylines and polygons) and *box* (axis-parallel rectangle in 2D space) have been realized in the system with corresponding geometric functions like *intersect, inside, distance* etc.. In GEO^{++} additionally the 2D data types POINT2, POLYLINE2 and POLYGON2 are integrated[1]. The geometric functions of GEO^{++} can be subdivided according to the data types of their results into the following four groups. They can be extended at any time:

---

[1]. The Postgres type *path* is not used in GEO++.

1. *boolean values as result*
   (e.g. equal2PntPnt, contain2PgnPgn);
2. *Atomic geometric objects as result*
   (e.g. gravCenter2Pgn, convexHull2Pnts);
3. *Complex geometric objects as result*
   (e.g. inter2PgnPgn, delaunay2Pnts);
4. *Scalar values as result*
   (e.g. distance2PntPnt, area2Pgn).

GEO^{++} is implemented in C^{++}. The most important functionality is the following: as could be expected, in GEO^{++} Postgres relations can be read in a browser and the tuples can be edited. Especially operators and user-defined data types can be inspired and the *where* restriction in SQL query expressions can be shown graphically and syntactically checked via the Postgres system catalogue (meta relations) in a tree-like sturcture. Geometric results of database queries (points, lines and areas) can be visualized on a geographical map inclusive of labels for the objects. Furthermore, graphical objects can be selected with the mouse to get detailed information of Postgres tuples. Attributes for point, line and area data can be generated and updated in an editor.

The graphical functionality of GEO^{++} includes the editing of objects and functions like zooming. They are supported by the R-Tree and a multi user access is possible. With the *Postgres trigger* the graphical display of relations can be redrawn automatically. This means that dynamic map displays can also be generated with moving objects.

Geometric operations are directly defined in Postgres. The visualization, however, is realized by new user-defined GEO^{++} *query shapes*. In such a query shape at least one *draw* method and one *distance* method has to be implemented which are needed for the drawing of the query shape on the map and for the realization of the selection operation.

Besides vector data in GEO^{++} raster data can also be processed. For that reason GEO^{++} has been extended by a special *query shape* that graphically displays a raster data type. The raster data are stored in clusters of variable size with Postgres tuples. Again the R-Tree is used for that purpose. As known from other GIS a scanned raster map can be taken as background of a digitalized vector map.

Another advantage of GEO^{++} is that advanced functions like the shortest path function can be implemented in the system to realize the efficient search in a graph. This is done by the concept of *views* as they are well known from relational database systems technology. The direct access to the tuple of the database is provided. Although the functionality of GEO^{++} is not as advanced as that of commercial GIS the principle of extensibility in the system architecture has been revolutionary for the development of (commercial) GIS.

3.2.2 GeO$_2$

Like GEO^{++} GeO$_2$ is not a complete GIS, but a research protoype with main emphasis on object-oriented modelling and the efficient management of geo-objects. The system is based

Chapter 3. Examples of Today's Geoinformation Systems 51

upon the OODBMS O_2 (DEUX 1990). GeO_2 distinguishes itself from other GIS by the fact that extensions like spatial access methods can be implemented at deep system levels. The O_2-Tools provide interfaces in C, C^{++}, O_2C, O_2SQL and O_2Look. They support the programming, descriptive queries and query by example, respectively.

GeO_2 can also be used as an experimental tool for GIS developers especially for performance tests of spatial access methods, map overlaying algorithms and spatial query languages (DAVID et al. 1993).

The conceptual data model of GeO_2 is subdivided into two levels. The "semantic geographical data model" at the first level is an entity relationship model that is extended by the two concepts of inheritance and propagation. The "localization data model" at the second level is equivalent to the abstract data type (ADT) "geometry". This ADT is function-based, i.e. it does not only use predicates but functions. For example, geometric operators like the intersection of two objects can be defined. This advanced operator returns a geometry. The intersection of a line with two areas can be computed in the following steps. First the intersection of the two areas is computed and after that the result is intersected with the line.

Each entity that is defined in the semantic data model can have one or more attributes of the type *geometry*. As usual with object-oriented DBMS the semantic schema of the users is translated into classes of the OODBMS. The geometry ADT is consequently implemented as geometry class that consists of geometric and topological primitives. We will introduce these primitives below.

For the efficient management of the geometry and spatial relationships between the entities in GeO_2 three levels of spatial representations are implemented. The most simple representation is the spaghetti model. In this model the geometric primitives (point, line, polygon) are independent of each other without any topological relationship. At the medium level a non--planar graph is defined which could support the route search in an underground network. The elementary classes are directed edges *(DartPos, DartNeg)* which inherit from the dart class as well as nodes. In a winged-edge representation the references from the current edge to the preceding and the following edges are stored. Areas are implemented with their contour in the two classes *SimpArea* and *CompArea* which stand for simple areas and for areas with holes, respectively. The *SimpArea* and *CompArea* classes inherit from the *SimpPoly* and the *CompPoly* class. Finally at the top level the map level is designed that corresponds to a topological map model with nodes and edges of a planar non-connecting graph with additional areas. Furthermore, in the graph the relationship between an area and its connected components (holes) is managed. In such a model neighoured areas and the sharing of boundaries are particularly supported. The physical spatial clustering of geo-objects is not provided by GeO_2.

3.2.3 GODOT

GODOT (EBBINGHAUS et al. 1994) is a prototype of an open object-oriented geoinformation system with client/server architecture based upon the OODBMS ObjectStore. Therefore ObjectStore has been extended by geospecific data types and operators. As user-interface in GODOT an interactive interface on the basis of XWindows - OSF/Motif, a UNIX command interface and a C/C++ programming interface is provided. The architecture of GODOT consists of the GODOT kernel system and of different extensible components like the query component and the graphics component etc.. The kernel system is directly based on ObjectStore and is implemented in C++. It includes implementations for base classes of geo-objects. Furthermore, the kernel system consists of a transaction management and spatial clustering of data with z-values, i.e. quadtree technology. The clustering guarantees an efficient preselection during spatial queries.

In the GODOT query component the query language of ObjectStore is extended by GIS specific geometric and topological predicates. That is why the query component in principle can be steered via all of the three interfaces, i.e. interactively, with UNIX commands and with the query language which is embedded into C/C++. With the graphics component geo-objects can be selected graphically and the results of queries from the query component can be visualized graphically.

In GODOT there is a general class for thematic objects from which all other geo-objects are derived. The object model of GODOT coarsely is organized into three classes of geo-objects for thematic objects, geometric objects and representation objects. Geo-objects can be composed by one or more thematic objects (part_of and parts relationship). So called meta objects can be managed in a data dictionary by using the object browser of ObjectStore.

In GODOT to each geo-object a set of geometric objects can be referenced. This means that such geo-objects can be defined as those that have non-connecting geometries or geometries of different geometric types like *regions, arcs* or *points*. The topological relationships between the geometric objects are explicitly managed with object pointers.

The representation objects, especially the cartographic objects, i.e. instances of the class *map* describe the graphical representation of thematic objects. Examples are the thickness of a line or the colour of an area. Cartographic objects that thematically fit together can be composed in one layer. It is possible to use a layer for more than one map. The cartographic representation objects refer to their corresponding geo-objects via reference objects. Furthermore, pointers to an explicit representation geometry are managed which mostly differ from the real geometry of the object, e.g. because a generalization of the object has taken place.

One of the test applications for GODOT has been realized on the basis of ATKIS data (AdV 1989), the official topographic cartographic information system of the study group for the geodetic admininistrations of the Federal Republic of Germany (EBBINGHAUS et al. 1994).

3.2.4 GeoStore

As the last research prototype in this list we briefly introduce GeoStore that has been developed in the Collaborative Research Centre 350[1] at Bonn University[2]. GeoStore is the result of a close cooperation between the Institute of Computer Science III and the Geological Institute at Bonn University.

---

[1] Collaborative Research Centre 350 "Interactions between and Modelling of Continental Geo-Systems" at Bonn University.
[2] Working groups of Armin B. Cremers and Agemar Siehl.

Chapter 3. Examples of Today's Geoinformation Systems 53

GeoStore (BODE et al. 1994) is a research prototype which is based upon the OODBMS ObjectStore to manage geologically defined geometries. The implementation is concentrated on the special 3D/4D database requirements of geology that emerged from the interactive geological modelling of the Lower Rhine Basin, Germany. The geological modelling process is characterized by three steps: in the first step line data are generated from point sets of digitalized x-sections. In the second step triangulated areas are generated between these sections (fault surfaces and stratigraphic surfaces). Finally in the third step volumes are constructed from the surfaces by an interactive adaptation of the sections, fault and stratigraphic surfaces.

Figure 3.7 shows a part of the geological object model that has been developed between computer scientists and geologists. The most important classes are derived from the relevant aspects of the interactive geological modelling in the examination area of the Lower Rhine Basin, Germany. These classes are *Section, Stratum* and *Fault*. The geometry of a stratum or a fault can be modelled as a stratigraphic line *(StratLine)* or as a fault line *(FaultLine)* on a section as well as a stratigraphic surface *(Stratum)* or a fault surface *(Fault)* between two sections. Surfaces are represented as extended simplicial 2-complexes (triangle networks in 3D space) which additionally represented topological information. The class *StratInSection* has been introduced to realize a fast access from stratigraphic lines to the corresponding section. A *GeoPoint* is a point with additional geo-attributes like "name" or "lithology".

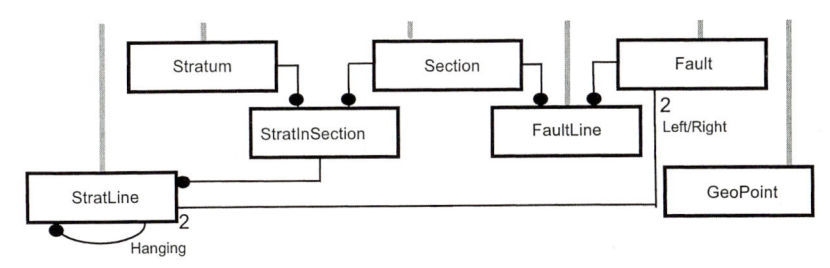

Fig. 3.7 Part of the GeoStore object model

The database functionality of GeoStore can be subdivided into the following types of queries:

1. *The browsing of geo-objects;*
2. *Queries on point sets and line data;*
3. *Queries on surface data including geometric operations;*
4. *2D- or 3D/4D visualization of geo-objects and query results.*

The results of spatial queries like the intersection of a horizontal or vertical plane with surfaces or the intersection between two surfaces in 3D space can be visualized with a 2D or 3D visualization. Afterwards additional information can be obtained by navigating in the database browser. Figure 3.8 shows a 3D visualization of some horizons from the so called Erft Block of the Lower Rhine Basin which have been selected by the database.

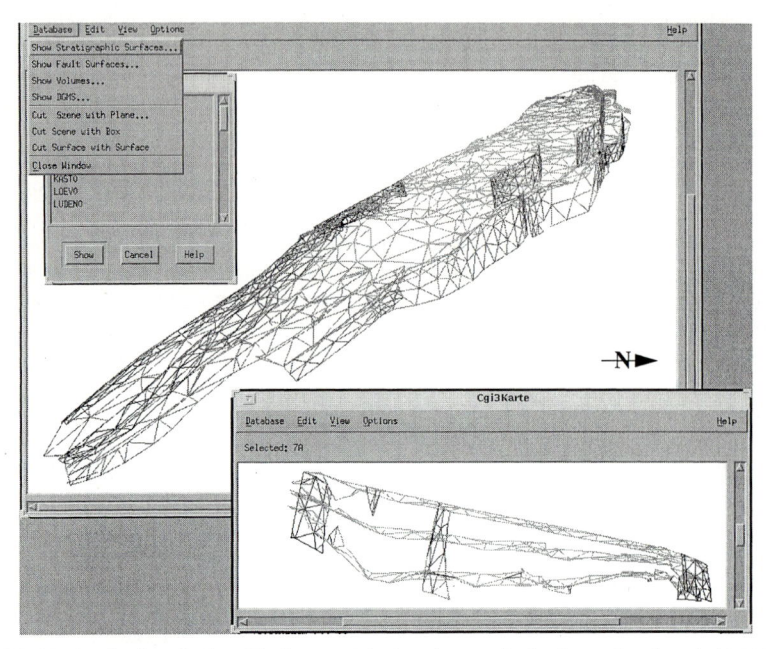

Fig. 3.8 3D visualization of selected horizons and database browser in GeoStore taken from the Erft Block of the Lower Rhine Basin, Germany.

In Fig. 3.8 the "switching" between the database browser and the visualization is to be seen. GeoStore originally has been applied only to the examination of the Lower Rhine Basin, Germany. However, other examination areas could be integrated into GeoStore on top of the object-oriented geodatabase kernel system GeoToolKit (Balovnev et al. 1997a). The object-oriented toolkit approach on the basis of an OODBMS can be seen as an important step on the way to open geoinformation systems.

3.3 Comparison

Today's geoinformation systems have closed system architectures and their internal system functions are difficult to extend. The spatial data handling of today's GIS usually is implemented upon internal representations which cannot be accessed from external software systems. Often only the thematic, i.e. non-spatial data are managed by a database management system (DBMS), which enables the access from outside the GIS. This fact, however, is in conflict with the interdisciplinary character of geo-applications. That is why the formerly closed geoinformation systems have to open their architectures, i.e. especially their data structures and access methods should be usable from outside the GIS.

In research the number of geoinformation systems based upon object-oriented technology is growing in comparison to the former layer-based systems (GÜNTHER and LAMBERTS

1992; VIJLBRIEF and OOSTEROM 1992; DAVID et al. 1993; BODE et al. 1994; BECKER et al. 1996; BALOVNEV et al. 1997a). Seen from a computer science perspective besides object-oriented modelling techniques, modern database technology takes a key position for the design and realization of GIS. The database management system guarantees the consistency of the data source, the protection against data losses (recovery), the support of multi user transactions (concurrency control) and a high data integration, i.e. avoidance of data redundancy. Object-oriented database technology also seems to be well suited for the integration of GIS. However, today there is not much experience with the modelling and management of geodata on top of OODBMS. Today's OODBMS are rather extensible developer toolkits than ready-to-use and simply usable query tools.

Previous geoinformation systems are predominantly restricted to the data processing of 2D data. Exceptions are additional modules, e.g. for the visualization of digital elevation models. Operations for the processing of 3D data are missing as well as the management for moving objects. However, for many open geoscientific problems in the different disciplines like geography, geology or geophysics the modelling, management and analysis of 3D/4D geodata are the precondition for their solution. If we consider the data management of today's GIS we realize that often not even the two-dimensional spatial data are managed by a DBMS. Instead the management of geometric data -mostly realized as maps- is directly done by the file system. This leads to two problems: first the objects have to be composed by the user a posteriori, because geometric and non-geometric data are managed separately. This leads to a lower performance for combined geometric and non-geometric database queries. Secondly, one has to give up the advantages of a DBMS for the management of large sets of geometric data. This means that, for example, the geometric data are exposed to a higher risk of data losses. But why are the geometric data of GIS not managed by the DBMS? One reason is that hitherto nearly all of today's database management systems only support one-dimensional indexes like the B-Tree (BAYER and MC CREIGHT 1972). More-dimensional access methods like the R-Tree (GUTTMAN 1984), the Grid File (NIEVERGELT et al. 1984), the LSD-Tree (HENRICH et al. 1989) or others (GAEDE and GÜNTHER 1998) are not yet provided[1].

---

[1]. First approaches are extra modules of RDBMS like the so called data-blades of Informix.

Chapter 4

Data Modelling and Management for 3D/4D Geoinformation Systems

This chapter goes into the spatial and temporal concepts of geoinformation systems and describes the modelling of 3D/4D objects. The specification of topological relationships for non-primitive geo-objects and temporal state transitions for topological relationships are consolidated. Basis operations for the mapping of state transitions for tetrahedra are given. Concerning the management of spatio-temporal objects the checking of integrity constraints for geological objects and the spatio-temporal database access for large object sets are discussed.

4.1 Relevance of Space and Time

4.1.1 Space

Usually we observe the world and also geoscientific phenomena in three-dimensional space. The spaces that are relevant for geoinformation systems can be seen on three different abstraction levels (BREUNIG 1996): the *topological space, the metrical space* and the *n-dimensional Euclidean space*.

As we know from mathematics a *topological space* can be defined with open point sets (QUERENBURG 1979). Two objects (point sets) belong to the same topology if their point sets are neighboured. The topological concepts of "boundary" and "interior" of an object abstract from a metrics and from Euclidean coodinates. Thus topological relationships between objects can be specified without knowledge of the geometry. Topological spaces are the first and top abstraction level of the mentioned spaces.

On the second abstraction level we can define metric spaces. They are an important class of topological spaces. M is a *metric space,* if a mapping d: $M \times M \to R_0^+$ exists (so called metrics on M) with the following properties $\forall\ x, y, z \in M$:

(1) $d(x, y) = 0 \Leftrightarrow x = y$ (identity axiom),
(2) $d(x, y) = d(y, x)$ (symmetry axiom),
(3) $d(x, z) \leq d(x, y) + d(y, z)$ (triangle inequality).

The distance *d(x, y)* between *x* and *y* is a metrics on the topology. For each topology we can define different metrics like the minimal distance, the minimal maximal distance or the

Hausdorff distance (GRÜNBAUM 1967). Besides the topology the introduction of the distance is an important precondition for the orientation in space within geoinformation systems.

Finally on the third abstraction level the *n-dimensional Euclidean space* is defined. Euclidean space is directly based on the cartesic coordinate system. For geoinformation systems the two- and three-dimensional Euclidean space are of special interest. A special motivation for a three-dimensional modelling in geo-applications is that the representation of the Earth's surface (2.5D, digital terrain model) is not sufficient for many applications. Classical examples are the deposit exploration and subsurface geology. But also for new applications above the Earth like environmental monitoring, city planning or mobile computing, 3D objects are necessary. They can be represented in a boundary model or in a full 3D volume model.

4.1.2 Time

The world is so organized that we can perceive it in space and time. Phemomena are running as events or processes that can spatially and temporally be observed. An observable geoscientific phenomenon can be examined with different methods, e.g. empirically with the generation of a time series or analytically with numerical modelling. In the GIS literature different types of time are distinguished like the "regional time", the "geoscientific time", the "registration time" and the "database transaction time" (WORBOYS 1992, 1995; BILL 1997; SPACCAPIETRA et al. 1998). The capture time and the database transaction time which describe the insertion and the existence interval of an object in the database respectively, can easily be modelled as an additional attribute of an object. Thus they can be stored in the database. More complicated, however, is the case with the geoscientific or real time. Only certain states can be stored in the database. Between them an interpolation has to be executed, e.g. to describe the animation of a geoscientific process.

4.1.3 Analogies between Space and Time

In space as well as in time we can define *topological and metric properties* that lead us from general and abstract views of topology to exact statements about distances between objects. Geometry refers to the *absolute space* and the *absolute time,* respectively. Against that with topology we can describe relative relationships between geo-objects in space and time *(relative space* and *relative time).* For example in geology the relative time scale is relevant, because in many cases there is no absolute time scale where the boundaries between geoscientific periods can be fixed.

4.1.3.1 Topological Properties

In the plane (2D) we can imagine a topological object as a "rubber sheet" (WORBOYS 1995) that can be stretched, but not torn or folded. Analogously, in three-dimensional space topological objects can be modelled with elastic material. For example, a sphere and a bottle have the same topology, because a bottle can be transformed into a sphere by turning out the interior side which is a pure topological operation.

By the *global topology* between several geo-objects we mean their spatial relationships between each other (inside, intersect etc.). Against that the *local topology* of a single geo-object consists of the spatial neighbourhoods of its sub-geometries, i.e. the connection between points, edges, surfaces and volumes of the objects (their discretization). We presuppose that each geo-object can be decomposed into corresponding primitive geometries.

Concerning *time* the global topology describes the temporal position between objects like the temporal neighbourhoods or intersecting time intervals. The *local topology* is the temporal change of the objects' discretization.

An Example: Topology in Geology

In many geo-applications geometry and topology in space and time are independent of each other, but not in geology. According to Walther's facies law (WALTHER 1893) the stratigraphic sequence and lithology of strata reflect the temporal trace of the geological processes of sedimentation which in many cases has its origin millions of years ago. The law says that in the geological past only such strata have been deposited on top of each other (vertical facies change) that today are sedimented beside each other (horizontal facies change). For a geologist it is obvious that both notions of topology in space and time belong together, i.e. they are strongly correlated: the temporal series of the strata is also the (recent) spatial series. The geometry (z-axis) and the time axis are strongly correlated, if not in a concrete case where geological faults or folds appear that could distort or even invert the time vector.

The topology of a geological stratum can be described in a topological graph (BOUILLE 1976). However, this only works as long as the series of strata are not too complex. A stratum in a graph can be represented as a non-intersecting closed curve, the so called Jordan curve. The topological relationship between a stratum and other strata can be defined with *Jordan's curve theorem for simple closed curves (SCC)*:

The complement of a SCC is not connected, but subdivided into two connected components. One of them is bounded (the interior of the SCC) and the other is not bounded (i.e. it has no boundary and is the exterior of the SCC).

4.1.3.2 Metric Properties

The spatial distance between two extended geo-objects is usually defined by distance measures like the minimal, the maximal or the Hausdorff distance (GRÜNBAUM 1967).

Such an absolute distance is not existing in time. We at most can give, e.g. a (travel) time, which gives the time under certain conditions we need to travel from a location A to a location B. In WORBOYS (1995) it is shown that the "travel time topology" (see chapter 2.1.1.3) is a to-pological space T consisting of a collection of sub-sets of T, the so called neighbourhoods that meet the following two conditions[1]:

---

[1]. A restriction is that the speed has to be constant.

1. Each point in *T* is in a neighbourhood;
2. The intersection of two arbitrary neighbourhoods of a point *x* in *T* includes the neighbourhhood of *x*.

Topological and Geometric Neighbourhood

Neighbourhoods of objects in space and time can be described topologically as well as geometrically. In the first case the neighbourhood is, e.g., described in a graph with shared edges between two nodes. This kind of neighbourhood plays an important role for an efficient realization of geometric algorithms on complex geo-objects like the interpolations on triangle networks. Against that in the second case, the geometric description, also the measure of the distance decides about the neighbourhood. This kind of neighbourhood is relevant for the physical spatial clustering of geodata in spatial and temporal access methods.

4.1.3.3 Field-Based and Object-Based Modelling

In chapter 2.1.3 we have already shown that the field-based and the object-based approach of spatial data modelling can be extended for time. The two-dimensional raster in space corresponds to a scene in time. The scene represents a raster with a given time. Time is a function of each raster cell in different times or time intervals. However, it is not possible to differentiate the single objects of a scene. This would be interesting, because a temporal refinement of a geoprocess could lead to detailed knowledge comparable to the knowledge gained from a spatial refinement of certain regions like mountains in a digital elevation model. Also hierarchical spatial and temporal approaches have to be considered (DE FLORIANI et al. 1994; GRIEBEL 1994). The field-based approach in time can be implemented with time stamps. For each scene exactly one time stamp has to be referenced.

In the object-based approach each object can have its own time. The advantage is that the temporal development of single objects can be compared with each other. As the time stamp corresponds to separated objects, time can also be stopped for certain objects and at the same time run on for other objects. We will pick up the object-based approach for time modelling again in chapter 4.2.1.3.

4.1.3.4 Approximation in Space and Time

In two- or three-dimensional Euclidean space a continuous curve can linearly be approximated with a sequence of segments (polyline). A first approximation can start from segments that are oriented parallel to the coordinate axes. They build a step function. A more detailed approximation would avoid this restriction, i.e. arbitrarily oriented segments of a given length are given. Obviously the area under the curve can be approximated with the sum of the rectangles under the curve (integral) and the estimation can be improved, if we use triangles instead (Fig. 4.1a). The approach can be extended to estimate the volume under an area. Therefore multi-dimensional integrals are used, i.e. the cuboids and tetrahedra under the area are computed, respectively.

We assume a linear spatial approximation and presuppose that for each time t_i, ($i \in 0..n$) a computable linear spatial approximation exists for each spatio-temporal object O_i. Then we can easily approximate continuously changing objects, if we only allow a change of the geometry and topology for discrete times. Figure 4.1b shows the linear temporal approximation of a curve with a change of geometry and topology (in this case the connection of the points) for the times t_1, t_2 and t_3. If the number of base points for the times t_i and t_{i+1} are not identical, then we have to define a unique mapping of the points and their connections between two time steps (t_i, t_{i+1}), respectively.

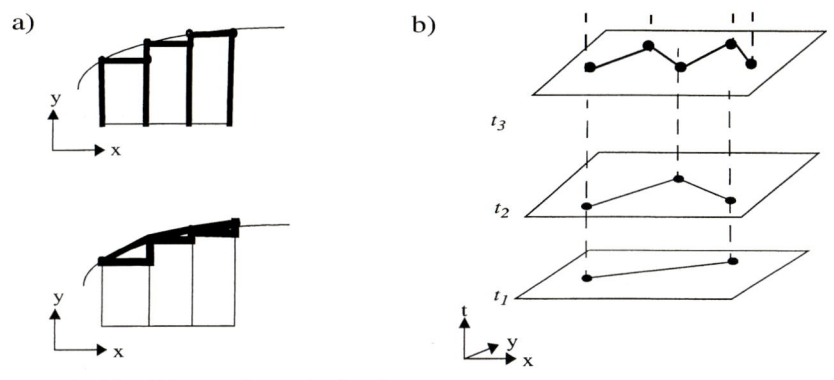

Fig. 4.1 a) Spatial and b) temporal approximation of a curve

The model of spatio-temporal information in the Euclidean plane of WORBOYS (1992) is an example for a model with change of the spatial extension for discrete times. The spatio--temporal objects are represented as prisms. The base area stands for the geometry and the z-axis is corresponding with time. Other authors like ERWIG et al. (1997b) have also used this representation for spatio-temporal objects. Figure 4.2 shows an example and the realization of Worboys' model with "ST-complexes".[1]

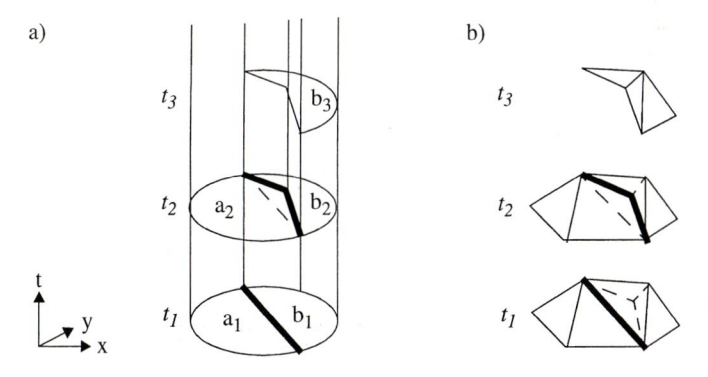

Fig. 4.2 a) Representation of discretely changing area objects as upright prisms
b) Realization as ST-complex

---

[1] Spatio-temporal complex.

The realization of a single ST-complex as has been implemented in WORBOYS (1992) has the disadvantage that for changes of the geometry also the topology of the current time and of all past times has to be updated a posteriori (see dashed lines in Fig. 4.2b). These updates are necessary, because the spatial conditions of a complex (see definition 2.2) have to be met at every time. As a consequence the discretization gets finer at every time step. This, however, usually leads to costly update operations in the database.

4.1.4 Differences

Obviously the Euclidean metrics which is defined with an algebraic operation (the scalar product) is defined in the Euclidean vector space. But there is no unique Euclidean metrics for both concepts, space and time. Thus space and time cannot always be seen symmetrically, i.e. the "fourth dimension" often has to be treated differently from the spatial dimensions. The temporal dimension has no metrics that is corresponding to the spatial metrics. This fact has to do with the concept of speed. Corresponding to a spatial metrics, each constant speed of an object leads to another angle between the coordinate axis; i.e. every object will have its own temporal and local coordinate system. Temporal metrics are object-based and have to be seen relatively to other objects. Spatial metrics, however, can be seen absolutely in space with one coordinate sytems for all objects.

Algebraic properties of space lead to concrete, specific properties in the Euclidean vector space. They correspond directly to Euclidean coordinates. A temporal concept that corresponds to the Euclidean vector space, however, does not exist. Two equal distances in time can have two different meanings, because speed is involved. The consequence is that for different objects or different speeds different coordinate systems with different angles for the axes have also to be considered.

Finally we should notice that the units of measure for spatial and temporal systems are also different. For space one unique system is usually used, the decimal system (metre). For time, however, a different system of units is used like 7 days of a week, 24 hours of a day, 60 minutes of an hour and 60 seconds of a minute, respectively.

4.2 Modelling of Spatial and Temporal Objects

4.2.1 Suitable Spatial and Temporal Representations

The suitability of different spatial representations (REQUICHA and VOELCKER 1982; MÄNTYLA 1988) for geoscientific applications has been examined in great variety (PEUQUET 1984; ANDERL and SCHILLI 1988; RHIND and GREEN 1988; AdV 1989; JONES 1989; GUENTHER and BUCHMANN 1990; SAMET 1990; KRAAK and VERBREE 1992; MALLET 1992a; SIEHL 1993; BREUNIG 1996 etc.).

For objects that are represented in the boundary representation, detailed information exists about the geometry and topology of their hull. The volume can be computed with the hull information. Structural information (topology), however, and thematic information for differentiated attribute values are missing. That is why the boundary representation is no "real" 3D representation.

Against that the 3D representation of Computational Solid Geometry (CSG) is characterized by fixed geometries like cubes, cylinders, pyramids etc. and the logical operators union, intersection and difference between these primitives. With these operators, e.g. large sets of similar and industrially produced CAD objects can interactively be constructed. Objects are represented as trees. The interior nodes of the tree are the operators and the leaf nodes are corresponding to the geometric primitives. The CSG is well suited for the 3D representation of buildings, e.g. during the process of semi-automatic reconstruction from aerial pictures. In contrast to these "man-made objects", however, nature produces a lot of irregular geometries. That is why the CSG representation is often not well suited to represent geo-objects.

In the geosciences we need a simple, but easily configurable spatial representation. The octree (SAMET 1990) which is derived from the simple voxel representation subdivides objects into cubes with a given resolution. It also easily enables the interpolation of point attributes to the volume with its regular topological structure. The regular topology, however, also has disadvantages: the smallest cube determines the resolution of the whole octree. This leads to a waste of storage space and to an unnecessary representation of highly resolved regions for which, however, no detailed information is existing. Furthermore, in the geosciences the distribution of objects often is irregular so that regular partitions of space do not "fit" to the corresponding data points.

Latest developments show that in many geoscientific 3D applications irregular tesselation models are used (EDELSBRUNNER and MÜCKE 1994; BREUNIG 1996; CONREAUX et al. 1998; GERRITSEN 1998; MALLET 1998a,b). They are based on the theory of algebraic topology and characterized by a more detailed approximation of the geometry and a good adaptation to irregular distributions of data points. A disadvantage certainly is the high storage space needed. It can, however, be reduced by hierarchically organized tesselations (DE FLORIANI et al. 1994).

In the following chapter we introduce three spatial representations whose favourable properties are particularly useful for geo-objects.

4.2.1.1 Generalized Maps

The so called *generalized maps* are used in the GOCAD 3D modelling tool (LIENHARDT 1989, 1994; LIENHARDT et al. 1997; CONREAUX et al. 1998). In the GOCAD system the "G-Maps" have replaced the so called Weiler model (WEILER 1988) which is a boundary representation based on a topological connection of maps ("winged edge" representation). Generalized maps can be applied to 1D-3D geometries. The local topology between the points, edges and surfaces of an object is described by so called darts and by $\alpha 0$-, $\alpha 1$- and $\alpha 2$-connections, respectively (see Fig. 4.3). Furthermore, the external topology between an object and other objects can be described and the geometry of the objects including their properties is represented.

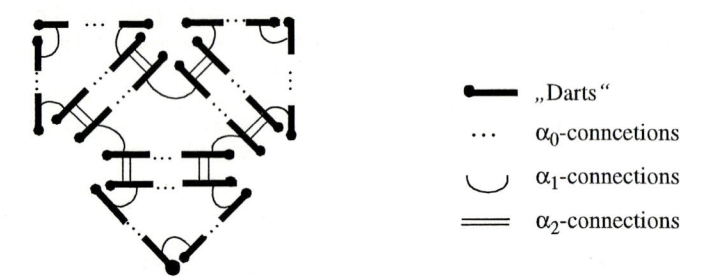

Fig. 4.3 Example of a G-Map

In the terminology of the "G-Maps" the local topology is called *subdivision* and the *connectivity* stands for the exterior topology[1]. The subdivision, e.g., describes the partition of a volume into tetrahedra and the partition of the single tetrahedra into its four surfaces. Furthermore, each surface is subdivided into its three adjacent edges and each edge consists of its two end points. Thus for a point its adjacent edges can easily be determined.

The connection describes, for example, the relationship of application specific geological objects like a stratum or a substratum to all its neighboured strata or substrata in a geological set of strata. The geometry includes the properties being attached to the local topology like single tetrahedra, triangles, edges and points that are denoted as the *embedding* of the object.

Another property of the generalized maps is that "non-manifolds"[2] can be modelled without the definition of new data types (LIENHARDT 1994). For a single embedding (frame) several *canvas* (interior topologies) can exist. Thus the concept of data encapsulation is applied. For example, the interior of a volume which is integrated in a geological set of volumes can be represented as a tetrahedron network, grid or parameterized volume. The links between frame and canvas propagate a change of the boundary of the geometry automatically into the interior of the object. Thus the consistency between frame and canvas is guaranteed.

We summarize the properties of generalized maps as follows:

1. The local topology of an object is described with links between geometric primitives (e.g. tetrahedra and their corresponding triangles, edges and points).
2. The external topology of an object is explicitly represented by a set of links to its neighbour objects (e.g. a set of double edges as link between the edges of two surfaces). Physically they only have to be stored once. The external topology also is used to describe topological relationships between geological objects like strata and faults.
3. For each object its interior topology (canvas) within a given boundary (frame) can be exchanged with other spatial representations.
4. It is also possible to represent non-manifolds.

---

[1]. In the GOCAD 3D modelling system (Tsurf company at Nancy, France) the implementation of a subdivision is called *canvas* and the implementation of a connection is called *frame*.
[2]. A non-manifold is characterized by the fact that in the neighbourhood of two neighboured interior points of an object there are points that are not belonging to the object.

However, the explicit representation of all links between all geometric primitives of the local and the global topology needs many pointers. This renders the mapping of generalized maps into a database more difficult, especially for large objects with several thousands of geometric primitives.

To the knowledge of the author, hitherto, temporal extensions for generalized maps have not been considered.

4.2.1.2 α-Shapes

An α-shape of a finite point set is a polytope[1] that is uniquely determined by a point set and a parameter α which controls the degree of detail for the α-shape (EDELSBRUNNER et al. 1983; EDELSBRUNNER and MÜCKE 1994). The α-shape intuitively describes the shape of the point set. A large α-value leads to a complete triangulation, i.e. the convex hull of the point set. A small α-value leads to a result near the original point set. An α-shape is defined as follows (EDELSBRUNNER and MÜCKE 1994; GERRITSEN 1998):

An α-complex is a finite simplicial k-complex. Let C_α be an α–complex and $\sigma_{yj}^{(k)} \in C_\alpha$ a k-simplex that belongs to C_α. Let $K_\delta^{(k)}(\sigma)$ be a k-dimensional sphere with radius δ, which contains in its boundary Bd $K_\delta^{(k)}(\sigma)$ exactly the k+1 edges x ($x \in y_j$). Notice[2] that $\sigma_{yj}^{(k)} =$ Conv y_j. Let $K_\delta^{(k)}(\sigma)$ contain no further edges, i.e. the interior Int $K_\delta^{(k)}(\sigma) = \emptyset$. Then we can define α-complex C_α. For its non-singular k-faces holds:

$$C_\alpha^{(k)} = \{\sigma_{yj}^{(k)} \mid \text{Int } K_\delta^{(k)}(\sigma) = \emptyset \text{ and } \delta < \alpha\}$$

α-shapes are variable simplicial complexes that are very useful for certain applications. Their construction, however, is costly (GERRITSEN 1998), because updates of the α-values lead to a new triangulation each time. The triangulation usually is computed in O(n*logn) time.

Additionally it is possible to attach weights at the points of an α-shape. The weighted α--shape W_α which is also called weighted α-complex (EDELSBRUNNER 1992) can be defined as the space of a sub-complex for a regular triangulation. In the first construction step the triangulation is computed and afterwards the specified simplexes are selected. The weighted α-complex is well suited for applications in which partial structures of triangulations are of special interest or for applications in which the shape of surfaces or solids changes in time. An example is the triangulated area representation of a meandering river network. The triangulate surface is not interesting for the geoscientist. The lines of the river network, however, are only a part of the triangulation and they are of high interest.

The α-shapes are also interesting, if the shapes of surfaces and solids are not exactly known and if they have to be determined experimentally by variable α-values. Obviously there are similarities to a constrain-controlled triangulation. However, during the generation of α-complexes no checking of the angles between the triangles takes place. Thus in some cases very thin triangles are generated which lead to computing inaccuracies especially in geomet-

---

[1]. In the literature of algebraic topology a polytope is the space taken as a basis for simplicial complexes.
[2]. Int stands for interior, Conv for convex hull.

ric algorithms. This is a decided disadvantage of α-complexes. Furthermore, an equally distributed data set is required as a framework for the α-complex.

4.2.1.3 Convex Simplicial Complexes

We have already introduced the definition of the simplicial complex in chapter 2.1.1.2. Simplicial complexes distinguish themselves for the modelling of geo-objects as follows:

1. A simple theory for the description of topological relationships;
2. The decomposition into simple geometric primitives (point, edge, triangle, tetrahedron);
3. A good approximation to the shape of "real objects";
4. The use of relatively simple interpolation and intersection algorithms that can be reduced to the single geometric primitives.

Simplicial complexes are a very suitable representation for irregular and complex, naturally shaped objects. Non-manifolds cannot explicitly be modelled, but they can be decomposed into several objects. For example, an object which is decomposed into k surfaces could be defined as a new user-defined data type.

In BREUNIG (1996) we have already shown the advantages of convex simplicial complexes. Their "bays" and "holes" have been filled with triangle and tetrahedron networks for surfaces and solids, respectively. These "virtual" triangles and tetrahedra have generated a convex shape of the surface and the solid, respectively. This leads to an advantage for the further processing of the objects especially in geometric algorithms: the algorithms are simpler and less special cases have to be considered.

Simplicial complexes can easily be extended for the representation of spatio-temporal objects by interpolating between the different internal states of the geometry (shape) and the topology (discretization of the triangles and tetrahedra, respectively). Such an extension is realized in the time model of the 3D modelling tool GRAPE (GRAPE 1997; POLTHIER and RUMPF 1995). For the interpolation in this model the precondition is required that the topology, i.e. the discretization like the number of triangles for triangle networks, must not be changed between two different states in time. For each time state (so called *TimeStep*) each object is represented with two simplicial complexes which have a different number of simplexes (so called discretization factor). With this trick -the change of the topology at defined times- the animation of objects can be executed in GRAPE. The first representation, the so called *post-discretization* corresponds to the approximation of the current state of the object. It has the topology that is needed for its current size and shape (Fig. 4.4). The second representation, the so called *pre-discretization* also corresponds with the approximation of the current state of the object. However, it has the topology in use in the previous state.

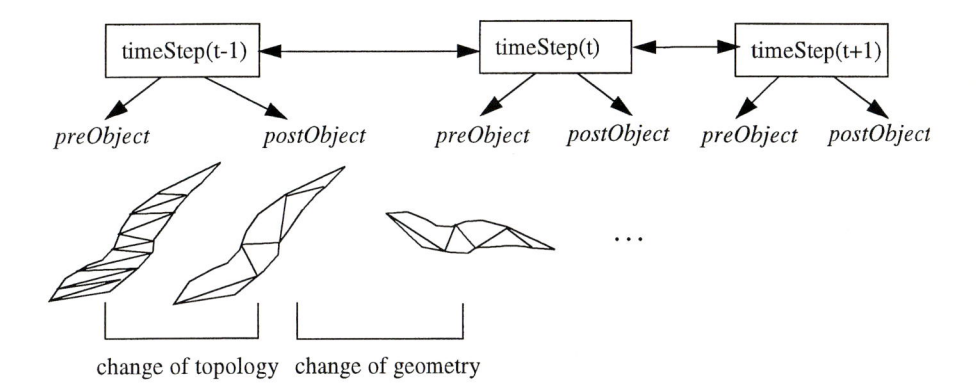

Fig. 4.4 Repesentation of simplicial complexes in time with interpolation between different time states

The interpolation between two simplicial complexes can always be done with the same discretization factor, i.e. with the same topology: the post-discretization of the previous state and the pre-discretization of the current state of the object (Fig. 4.4). Convex simplicial complexes can again be generated to compute the intersections between different geometries at certain times.

4.2.2 Spatial and Temporal Operations

4.2.2.1 Classification

Considering the examination of temporal relationships between objects by ALLEN (1983, 1984) and spatial relationships by EGENHOFER (1989) and KAINZ (1991) we can classify spatio-temporal operations uniquely according to their spatio-temporal types. We intend to use them as basis operations in a geodatabase. We subdivide the operations according to their resulting data type into *relationships, functions* and *operators*. Table 4.1 shows the types of spatio-temporal operations with the corresponding basis operations.

operation type \ spatio-temp type	topological	geometric	order
relationships: geoObject × geoObject → bool geoObject × geoObject 　　× vector → bool	disjoint; meet; overlap; covers/coveredBy; inside/contains; equal.		inFrontOf/ before; behind/ after.
functions: geoObject → real geoObject × geoObject → real		extension. distance.	
operators: geoObject → geoObject set of geoObject 　　→ set of geoObject* geoObject × geoObject 　　→ set of geoObject*	discretisize. interpolate.	 interpolate. intersection; union; difference.	

Table 4.1 Spatio-temporal basis operations for a geodatabase (geoObject* stands for a newly generated geo-object[1])

Relationships like the exclusion *(disjoint)* or the inclusion *(inside/contains)* evaluate a predicate, whereas *functions* return a real number as result. Finally operators return a geo-object or a set of newly generated, i.e. computed geo-objects. *Topological relationships* are invariant against geometric transformations, i.e. they stay unchanged if the coordinate system is translated, rotated or scaled. They describe the spatial position between geo-objects, e.g. if a well is inside a given region or if two geological faults are intersecting each other. The discretization (triangulation) is an operator that changes the topology of a single geo-object. The interpolation between a given set of points in the plane can be executed as a mapping on a (temporally changeable) line or an area. The classical *geometric set operators* are the *intersection*, the *union* and the *difference*. They return a set of newly generated geo-objects. *Met-*

---

[1]. New objects are not generated in any case. If the result consists of the empty set or exactly of the input objects, then no new objects have to be generated.

ric functions return the value of a geo-object's *extension* or of the *distance* between two locations. *Order relations* evaluate a predicate, i.e. they tell us for example if an object is spatially before or behind another or if it existed temporally before or after the other object.

4.2.2.2 Spatial Topological Relationships

Topological relationships have been identified as an important class of spatial relationships for geoinformation systems for several years (EGENHOFER 1989; KAINZ 1991; CLEMENTINI and DI FELICE 1994). The problem of former examinations, however, is that only for simple convex geometric objects and predominantly in 2D space, has an easily comprehensible set of topological relationships been defined for GIS users. Hitherto the definition of topological relationships for complex geometries has lead to many special cases with a large number of similar topological relationships (CLEMENTINI and DI FELICE 1994; EGENHOFER and FRANZOSA 1995; BREUNIG 1996). These relationships can no longer be classified by the 4-intersection method of EGENHOFER (1989). They often consist of a combination of other topological relationships. Figure 4.5 shows examples of such relationships between a line and an area in 2D space.

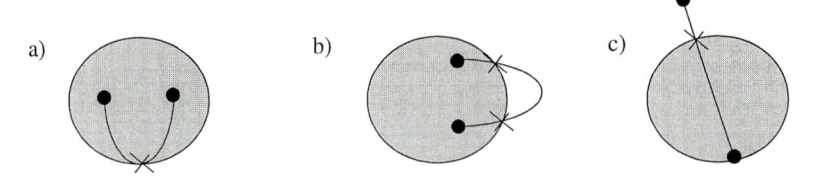

Fig. 4.5 Examples of topological relationships that cannot be classified with the 4-intersection method

The topological relationship of Fig. 4.5a intuitively corresponds to the *"covers"*-relationship, i.e. the touching of the line from the interior of the area. Then, by definition, also the boundary of the line intersects the interior of the area and the interior of the line must not intersect the boundary of the area. However, the interior of the line *is* intersecting the boundary of the area. This is true for both cases, if we define the boundary of the line as dependent on or independent of the space dimension in which the line is located (see Fig. 4.6). In Fig. 4.5b two intersecting points exist and in Fig. 4.5c an intersecting point as well as a touching point between the line and the disk exist. Thus both cases cannot be described with the 4-intersection method.

Fig. 4.6 a) Boundary of a line, defined for two-dimensional space (no interior)
b) Boundary and interior of a line independent of the dimension in space (boundary as end points)

In fig. 4.6a for the (poly-)line only a boundary, but no interior exists; i.e. lines are consi-dered in 2D space as "shrunk" areas that do not have an interior. Correspondingly in 3D space only solids have an interior. Lines and areas, however, do not have an interior. In Fig. 4.6b the boundary of the (poly-)line always consists of both end points of the (poly-)line, independently of the considered dimension in space. The interior in this case is the point set between the end points.

In Table 4.2 the definition of the *covers*-relationship as well as the result is shown with dimension dependent and dimension independent boundary @ and interior °.

line L /area A	@L ∩ @A	°L ∩ °A	@L ∩ °A	°L ∩ @A
definition of „classical" *covers*	¬∅	¬∅	¬∅	∅
covers with dimension depend. boundary and interior	¬∅	∅	¬∅	∅
covers with dimension independent boundary and interior	∅	¬∅	¬∅	¬∅

Table 4.2 Divergence of the "classical" covers-definition between a (poly-)line *L* and an area *A* (grey-filled fields)

As we have seen in Fig. 4.5a the definition of topological relationships causes problems, if:

1. The geometry of the objects goes beyond the primitives (point, segment, triangle or disk, tetrahedron or sphere),
2. Objects with different dimensions or a higher dimensional space are involved.

We introduce an approach for the definition of topological relationships between objects with complex geometries and with eventually different dimension or higher-dimensional spaces in 3D space. We solve the two mentioned problems as follows: each complex geometry, represented with a simplicial complex is decomposed into simplexes. The topological relationships are executed on the simplexes.

As an example we again take the topological relationship from Fig. 4.5a and suppose that the line and the area are in the same plane. The polyline is decomposed into two line segments (1-simplexes) so that the boundary (end point of the first and start point of the second, respectively) of the line segment touches the boundary of the disk -which is also approximated with 1-simplexes- from inside. Then the intersection between the boundaries and the interiors of the line segments and the area exactly corresponds to the "classical" *covers*-relationship, i.e. the values of line 1 in Table 4.2. Thus the topological relationship has been "broken down" on the simplexes (BREUNIG 1996).

Assuming that in the shown example the line and the area are not in the same plane, the line touches the area from outside and the intersection between the boundaries and the interiors of the line segments and the area corresponds to the "classical" *meet*-relationship.

Now we can describe topological relationships between simple objects with different dimensions and in a higher dimensional space. So we can solve the problem of defining topological binary relationships for objects with non-primitive geometry in 3D space. Therefore we represent the objects internally as simplicial complexes. We give rules that decide how the topological relationships between the single simplexes of two objects determine the topological relationship between the complete objects. We start with an example (Fig. 4.7) to illustrate the approach.

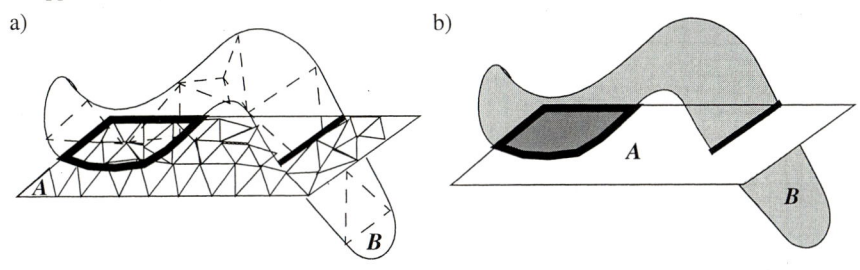

Fig. 4.7 a) Intersection and touching of the 2-simplexes for the surfaces A and B
b) A *overlap3D* B

The evaluation of the topological relationships between the 2-simplexes of the surfaces A and B in Fig. 4.7a lead to two observations:

1. ∃ a connected region in which 2-simplexes of A and B touch each other from outside *(meet3D)*;
2. ∃ a connected region in which 2-simplexes of A and B intersect each other *(overlap3D)*.

Connected regions are determined by the navigating on the neighboured simplexes. In the example we have presented a combined topological relationship *(meet3D* and *overlap3D)*. Thus it is not very useful to swamp the user with details like "*meet3D* takes place m times and *overlap3D* takes place n times". We propose to define exactly one *dominant topological relationship*. The detailed information can be given on demand. As we will see below the identification of an intersection and a touching in the example speaks for the decision to define *overlap3D* as the dominant topological relationship.

For the general case we give total orders to determine the dominant topological relationship (Fig. 4.8) from a set of combined topological relationships between simplexes S_A and S_B of two objects A and B:

(1) $(S_{Ai}\ overlap\ S_{Bj}) > (S_{Ai}\ covers\ S_{Bj}) > (S_{Ai}\ contains\ S_{Bj})$

(2) $(S_{Ai}\ overlap\ S_{Bj}) > (S_{Ai}\ coveredBy\ S_{Bj}) > (S_{Ai}\ inside\ S_{Bj})$

(3) $(S_{Ai}\ overlap\ S_{Bj}) > (S_{Ai}\ meet\ S_{Bj})$

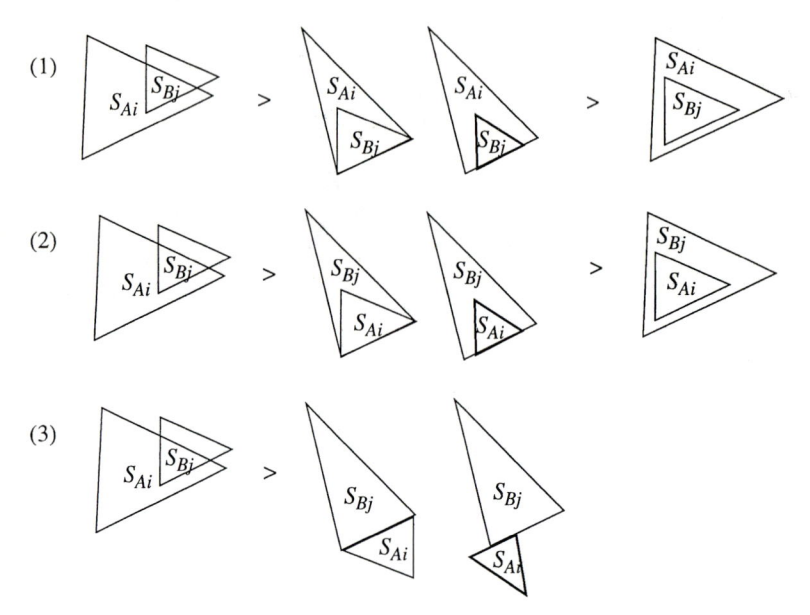

Fig. 4.8 Orders of topological relationships between simplexes S_{Ai} and S_{Bj} to determine the dominant relationship between two simplicial complexes A and B

If several of the topological relationships described above appear at the same time, the relationships which are standing on the left side of the >-sign "beat" the relationships on the right side. If, for example, the *overlap*-relationship between two simplexes appears, equally occurring *covers*- or *contains*-relationships between the other simplexes of A and B do not play any role, i.e. A overlap B is valid. In Fig. 4.9 we show further examples.

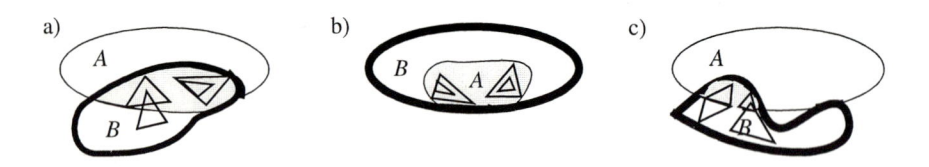

Fig. 4.9 Dominant relationships between two objects A and B which have been detemined with the order of topological relationships between single simplexes:
   a) $\exists\, S_{Ai}, S_{Bj}\,|\,(S_{Ai}\ overlap\ S_{Bj})\ \land\ (S_{Ai}\ covers\ S_{Bj})\ \land\ (S_{Ai}\ contains\ S_{Bj}) \Rightarrow A\ overlap\ B$
   b) $\exists\, S_{Ai}, S_{Bj}\,|\,(S_{Ai}\ coveredBy\ S_{Bj})\ \land\ (S_{Ai}\ inside\ S_{Bj}) \Rightarrow A\ coveredBy\ B$
   c) $\exists\, S_{Ai}, S_{Bj}\,|\,(S_{Ai}\ overlap\ S_{Bj})\ \land\ (S_{Ai}\ meet\ S_{Bj}) \Rightarrow A\ overlap\ B$

Notice that the combination of arbitrary other relationships with "disjoint" and "equal" never lead to a dominant "disjoint" and "equal" relationship, respectively. As a condition for A *dis-*

joint B holds: $\forall S_A \in A, S_B \in B: S_A$ *disjoint* S_B and for *A equal B:* $\forall S_A \in A, S_B \in B$ by pairs: S_A *equal* S_B.

We determine the minimal set of the "unique", i.e. not combined topological relationships between two simplicial complexes *A* and *B* as follows:

1. If all simplexes of object *A* are *disjoint* with all simplexes of object *B*
   $\Rightarrow$ *A disjoint B.*

2. If all simplexes of the objects *A* and *B* are by pairs *equal*
   $\Rightarrow$ *A equal B.*

3. If $\exists$ simplex of @*A* that touches a simplex of @*B* from outside *(meet)* and if all other simplexes of *A* are *disjoint* to all other simplexes of *B*:
   $\Rightarrow$ *A meet B.*

4. If $\exists$ simplex of *A* that intersects a simplex of *B (overlap)*
   $\Rightarrow$ *A overlap B.*

5. If $\exists$ simplex of @*B* that touches a simplex of @*A* from inside *(covers)* and if all other simplexes of *B* are *inside* @*A*
   $\Rightarrow$ *A covers B.*

6. If $\exists$ simplex of @*B* that touches a simplex of @*A* from outside *(coveredBy)* and if all other simplexes of *A* are *inside* @*B*
   $\Rightarrow$ *A coveredBy B.*

7. If all simplexes of *A* are *inside* @*B*
   $\Rightarrow$ *A inside B.*

8. If all simplexes of *B* are *inside* @*A*
   $\Rightarrow$ *A contains B.*

4.2.2.3 Temporal Change of the Spatial Topology

Going on another step we consider the spatial topology between geo-objects with their development in time. We then speak of *movements* and *changes of the extension* for geo-objects.

In the following graph (Fig. 4.10) the nodes stand for topological relationships and the edges for possible changes (see also EGENHOFER and AL-TAHA 1992; PAPADIAS et al. 1998). If we only allow movements of the objects (i.e. no changes of the extensions) in time, we get the relationship graph of Fig. 4.10. The edges between the relationships are marked with *m* for movement. This means that the change of the topological relationship in both directions of the arrow can only come about with a movement of the objects. We also assume that object *A* has been larger than object *B* from the beginning; i.e. the *equal*-relationship (*A = B*) cannot occur. *Equal* has to be seen as a special case of a topological relationship for objects with equal spatial extensions.

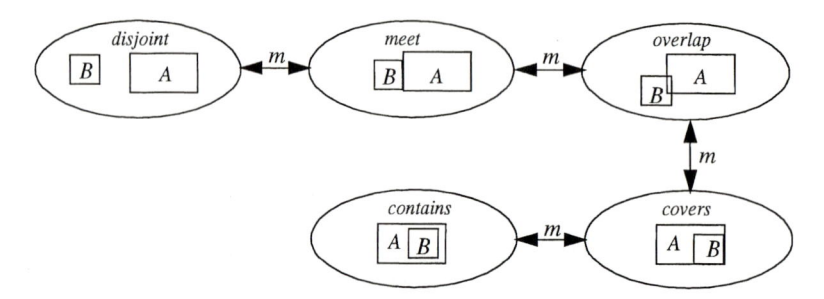

Fig. 4.10 Topological state transitions for moving objects A and B with constant extension[1]

Only the first of the complementary relationships *covers/coveredBy* and *contains/inside* can be valid, respectively. The reason for that is that we have assumed a constant size of the objects (no change of the extension) and we have assumed the size of A to be larger than that of B.

We get the graph of Fig. 4.11 for the topological state transitions, if we additionally allow that either the extensions of the objects are changing or the objects are moving in time. The edges of the graph are marked with *m* for movement and with *e* for extension. The double marked edge *m,e* has to be read as "the state transition of the topological relationship occurred either by the movement *m* or by the extension *e*". Now the objects A and B are also allowed to have the same size.

---

[1]. To simplify the presentation in the figure, only areas in two-dimensional space are shown. However, the same relationships are valid for solids in three-dimensional space.

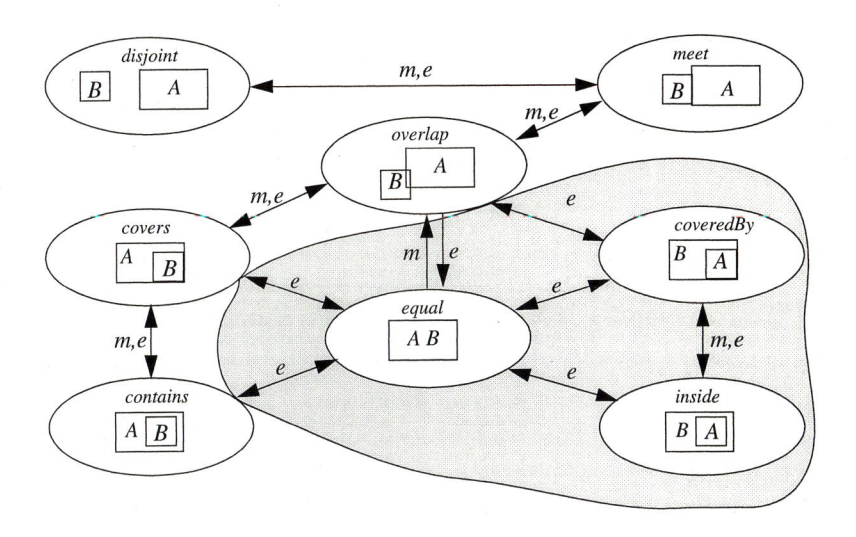

Fig. 4.11 Topological state transitions for moving objects A and B with variable extension

The relationships *equal, coveredBy* and *inside* (see the grey area in Fig. 4.11) are new and caused by the variable extension of the objects. They did not occur in Fig. 4.10. Furthermore, star-like arrows start from the *equal*-relationship to all relationships except to *meet* and *disjoint*. If we admit either a movement or a change of the extension, the transition from equal to overlap can only be caused by a movement *m*. In the other direction the transition can only be caused by a change of the extension *e* from *A* and *B*.

The topological state transitions can be specified as follows:

(1a) *disjoint(t_i)* → *meet(t_{i+1})*
  ⇔ ∃ simplex pair $(S_{Aj}, S_{Bj}) \in (A, B)$ | t_i: S_{Aj} *disjoint* S_{Bj} ∧ t_{i+1}: S_{Aj} *meet* S_{Bj}

(1b) *meet(t_i)* → *disjoint(t_{i+1})*
  ⇔ ∃ simplex pair $(S_{Aj}, S_{Bj}) \in (A, B)$ | t_i: S_{Aj} *meet* S_B j ∧ t_{i+1}: S_{Aj} *disjoint* S_{Bj}

(2a) *meet(t_i)* → *overlap(t_{i+1})*
  ⇔ ∃ simplex pair $(S_{Aj}, S_{Bj}) \in (A, B)$ | t_i: S_{Aj} *meet* S_{Bj}
  ∧ ∃ simplex pair $(S_{Aj+1}, S_{Bj+1}) \in (A, B)$ | t_{i+1}: S_{Aj+1} *overlap* S_{Bj+1}

(2b) *overlap(t_i)* → *meet(t_{i+1})*
  ⇔ ∃ simplex pair $(S_{Aj}, S_{Bj}) \in (A, B)$ | t_i: S_{Aj} *overlap* S_{Bj}
  ∧ ∃ simplex pair $(S_{Aj+1}, S_{Bj+1}) \in (A, B)$ | t_{i+1}: S_{Aj+1} *meet* S_{Bj+1}

(3a) $overlap(t_i) \rightarrow covers(t_{i+1})$
  $\Leftrightarrow \exists$ simplex pair $(S_{Aj}, S_{Bj}) \in (A, B) \mid t_i: S_{Aj}\ overlap\ S_{Bj}$
  $\wedge\ \exists$ simplex $S_{Bj+1} \in B \mid t_{i+1}: @A\ covers\ S_{Bj+1}$

(3b) $covers(t_i) \rightarrow overlap(t_{i+1})$
  $\Leftrightarrow \exists$ simplex pair $(S_{Aj}, S_{Bj}) \in (A, B) \mid t_i: S_{Aj}\ covers\ S_{Bj}$
  $\wedge\ \exists$ simplex $S_{Bj+1} \in B \mid t_{i+1}: @A\ overlap\ S_{Bj+1}$

(4a) $overlap(t_i) \rightarrow coveredBy(t_{i+1})$
  $\Leftrightarrow \exists$ simplex pair $(S_{Aj}, S_{Bj}) \in (A, B) \mid t_i: S_{Aj}\ overlap\ S_{Bj}$
  $\wedge\ \exists$ simplex $S_{Aj+1} \in A \mid t_{i+1}: S_{Aj+1}\ coveredBy\ @B$

(4b) $coveredBy(t_i) \rightarrow overlap(t_{i+1})$
  $\Leftrightarrow \exists$ simplex $S_{Aj} \in (A, B) \mid t_i: S_{Aj}\ coveredBy\ @B$
  $\wedge\ \exists$ simplex pair $(S_{Aj+1}, S_{Bj+1}) \in (A, B) \mid t_{i+1}: S_{Aj+1}\ overlap\ S_{Bj+1}$

(5a) $overlap(t_i) \rightarrow equal(t_{i+1})$
  $\Leftrightarrow \exists$ simplex pair $(S_{Aj}, S_{Bj}) \in (A, B) \mid t_i: S_{Aj}\ overlap\ S_{Bj}$
  $\wedge\ \forall$ simplex pairs (S_{Aj+1}, S_{Bj+1}) by pairs $\in (A, B) \mid t_{i+1}: S_{Aj+1}\ equal\ S_{Bj+1}$

(5b) $equal(t_i) \rightarrow overlap(t_{i+1})$
  $\Leftrightarrow \forall$ simplex pairs (S_{Aj}, S_{Bj}) by pairs $\in (A, B) \mid t_i: S_{Aj}\ equal\ S_{Bj}$
  $\wedge\ \exists$ simplex pair $(S_{Aj+1}, S_{Bj+1}) \in (A, B) \mid t_{i+1}: S_{Aj+1}\ overlap\ S_{Bj+1}$

(6a) $equal(t_i) \rightarrow covers(t_{i+1})$
  $\Leftrightarrow \forall$ simplex pairs (S_{Aj}, S_{Bj}) by pairs $\in (A, B) \mid t_i: S_{Aj}\ equal\ S_{Bj}$
  $\wedge\ \exists$ simplex $S_{Bj+1} \in B \mid t_{i+1}: @A\ covers\ S_{Bj+1}$

(6b) $covers(t_i) \rightarrow equal(t_{i+1})$
  $\Leftrightarrow \exists$ simplex $S_{B1} \in B \mid t_i: @A\ covers\ S_{Bj}$
  $\wedge\ \forall$ simplex pairs (S_{Aj+1}, S_{Bj+1}) by pairs $\in (A, B) \mid t_{i+1}: S_{Aj+1}\ equal\ S_{Bj+1}$

(7a) $equal(t_i) \rightarrow coveredBy(t_{i+1})$
  $\Leftrightarrow \forall$ simplex pairs (S_{Aj}, S_{Bj}) by pairs $\in (A, B) \mid t_i: S_{Aj}\ equal\ S_{Bj}$
  $\wedge\ \exists$ simplex $S_{Aj+1} \in A \mid t_{i+1}: S_{Aj+1}\ coveredBy\ @B$

(7b) $coveredBy(t_i) \rightarrow equal(t_{i+1})$
  $\Leftrightarrow \exists$ simplex $S_{Aj} \in A \mid t_i: S_A\ coveredBy\ @B$
  $\wedge\ \forall$ simplex pairs (S_{Aj+1}, S_{Bj+1}) by pairs $\in (A, B) \mid t_{i+1}: S_{Aj+1}\ equal\ S_{Bj+1}$

(8a) $equal(t_i) \rightarrow contains(t_{i+1})$
  $\Leftrightarrow \forall$ simplex pairs (S_{Aj}, S_{Bj}) by pairs $\in (A, B) \mid t_i: S_{Aj}\ equal\ S_{Bj}$
  $\wedge\ \forall$ simplex e $S_{Bj+1} \in B \mid t_{i+1}: @A\ contains\ S_{Bj+1}$

(8b) $contains(t_i) \rightarrow equal(t_{i+1})$
  $\Leftrightarrow \forall$ simplex $e\ S_{Bj} \in B\ |\ t_i$: @A $contains\ S_{Bj}$
  $\wedge\ \forall$ simplex pairs (S_{Aj+1}, S_{Bj+1}) by pairs $\in (A, B)\ |\ t_{i+1}$: $S_{Aj+1}\ equal\ S_{Bj+1}$

(9a) $equal(t_i) \rightarrow inside(t_{i+1})$
  $\Leftrightarrow \forall$ simplex pairs (S_{Aj}, S_{Bj}) by pairs $\in (A, B)\ |\ t_i$: $S_{Aj}\ equal\ S_{Bj}$
  $\wedge\ \forall$ simplexes $S_{Aj+1} \in A\ |\ t_{i+1}$: $S_{Aj+1}\ inside$ @B

(9b) $inside(t_i) \rightarrow equal(t_{i+1})$
  $\Leftrightarrow \forall$ simplexes $S_{Aj} \in A\ |\ t_i$: $S_{Aj}\ inside$ @B
  $\wedge\ \forall$ simplex pairs (S_{Aj+1}, S_{Bj+1}) by pairs $\in (A, B)\ |\ t_{i+1}$: $S_{Aj+1}\ equal\ S_{Bj+1}$

(10a) $covers(t_i) \rightarrow contains(t_{i+1})$
  $\Leftrightarrow \exists$ simplex $S_{Bj} \in B\ |\ t_i$: @A $covers\ S_{Bj}$
  $\wedge\ \forall$ simplexes $S_{Bj+1} \in B\ |\ t_{i+1}$: @A $contains\ S_{Bj+1}$

(10b) $contains(t_i) \rightarrow covers(t_{i+1})$
  $\Leftrightarrow \forall$ simplexes $S_{Bj} \in B\ |\ t_i$: @A $contains\ S_{Bj}$
  $\wedge\ \exists$ simplex $S_{Bj+1} \in$ @A $|\ t_{i+1}$: @A $covers\ S_{Bj+1}$

(11a) $coveredBy(t_i) \rightarrow inside(t_{i+1})$
  $\Leftrightarrow \exists$ simplex $S_{Aj} \in A\ |\ t_i$: $S_{Aj}\ coveredBy$ @B
  $\wedge\ \forall$ simplexes $S_{Aj+1} \in A\ |\ t_{i+1}$: $S_{Aj+1}\ inside$ @B

(11b) $inside(t_i) \rightarrow coveredBy(t_{i+1})$
  $\Leftrightarrow \forall$ simplexes $S_{Aj} \in A\ |\ t_i$: $S_{Aj}\ inside$ @B
  $\wedge\ \exists$ simplex $S_{Bj+1} \in$ @B $|\ t_{i+1}$: $S_{Aj+1}\ coveredBy$ @B

4.2.2.4 Change of Geometry, Topology and Connected Components for Objects in Time

We have already introduced the storage of changes for geometry and topology of geo-objects in discrete times in chapter 4.1.3.4. The subsequent change of the topology for all previous times in the database, however, is not acceptable because of the high costs for these update operations. Therefore we are searching for a way to approximate the geometry and topology changes in time so that a temporal sequence of geo-objects can efficiently be stored in a database. In this context geometry and topology can also be seen as the "shape" and the "structure" of the object, respectively.

The following considerations are influenced by existing applications from the geosciences, especially from geology (SIEHL 1993). In these applications the objects are seen as changeable in time. We first look at different cases that later have to be modelled as a sequence of objects with interpolations between the different times. We assume that the topology of an object at a single time is represented by a simplicial complex. In our concrete case a solid is represented with a tetrahedron network (3-complex) in three-dimensional Euclidean space.

In Fig. 4.12 the three different types of operations (state transitions) are listed:

1 and 2: Continuously changing geometry in time
("4D" with constant topology);

3 and 4: Continuously changing topology in time
("4D" with constant geometry);

5: Continuously changing geometry and topology in time
("4D" without restrictions).

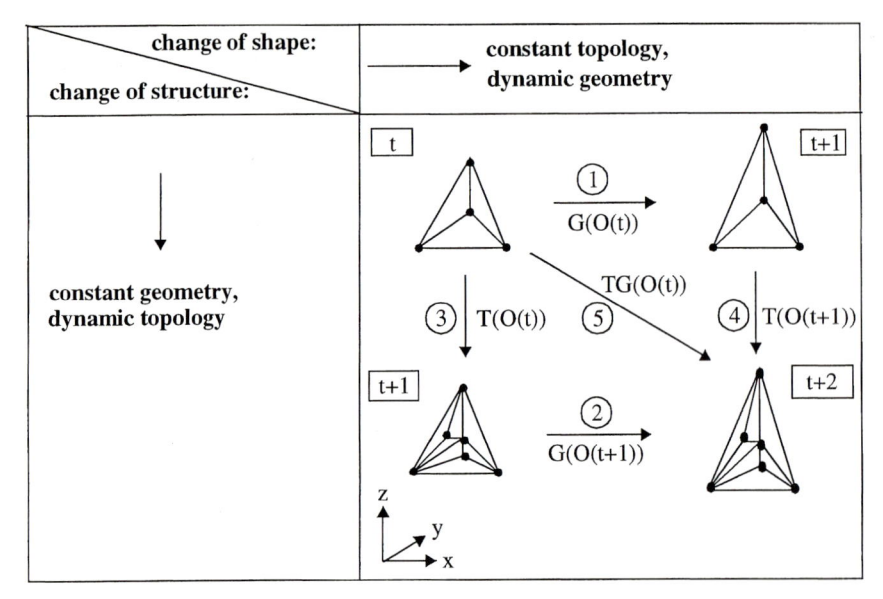

Fig. 4.12 Potentially same result for the change of the geometry or/and topology in time with tetrahedron networks

The changes of the geometry (1) and (2) are caused by a stretching of the respective geometry into the direction of the z-coordinate. Against that the changes of the topology (3) and (4) are caused by an insertion of a new point into the middle of the left triangle and into the triangle below in the respective object. For each inserted point a new tetrahedron is generated. In (5) the geometry changes and the topology changes are executed simultaneously.

It is interesting that the geometry change of an object in time $G(O(t))$ with the following topology change $T(O(t+1))$ potentially leads to the same result as a topology change $T(O(t))$ with following geometry change $G(O(t+1))$, i.e. for an adapted control of the discretization it holds that: $G(T(O(t))) = T(G(O(t)))$.

In chapter 5.2.7 we will see in an example that the temporal sequence of free shape deformations in geology can be constructed in a way that the geometry is changing between the different times, but topology is only allowed to change at certain states that have to be defined à priori. This means that the geometry is interpolated between the time steps and the topology

Chapter 4. Data Modelling and Management for 3D/4D Geoinformation Systems

is re-triangulated at certain states. However, this only includes the cases (1) and (2), because the topology between the time steps must not be changed. If we also allow the change of the topology (cases 3, 4 and 5), then mappings of the topology states have to be defined for all the time intervals $((t_i, t_i+1), i \in 1..n)$. For each tetrahedron we have to determine if it stays equal, if it is subdivided into several tetrahedra or if it is composed with other tetrahedra to a larger tetrahedron. Figure 4.13 shows structure, shape and component changing basis operations for tetrahedron networks. From these operations other complex operations can be composed. They change the geometry, topology and connected components of a tetrahedron network, respectively.

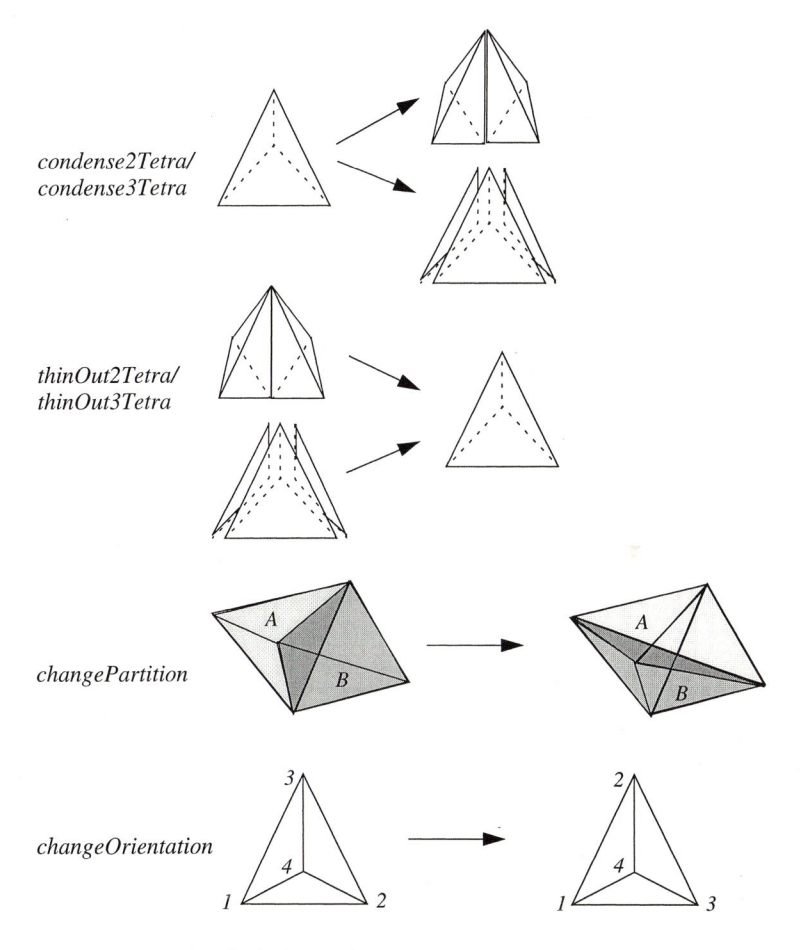

condense2Tetra/
condense3Tetra

thinOut2Tetra/
thinOut3Tetra

changePartition

changeOrientation

Fig. 4.13 Structure changing basis operations for tetrahedron networks

In the structure changing basis operations the topology of the tetrahedron network is changing by adding or deleting points. The discretization of a single tetrahedron is changing. Furthermore, the orientation of the points can be defined in different ways.

newTetrahedron

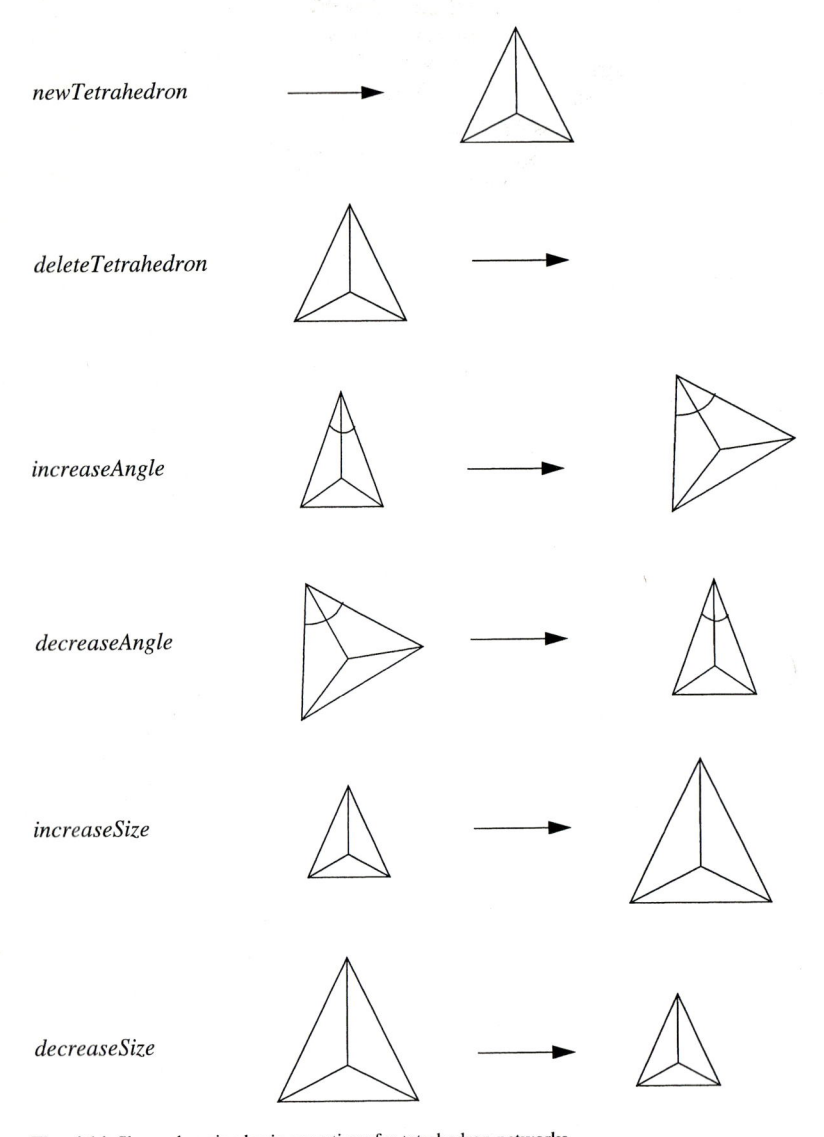

deleteTetrahedron

increaseAngle

decreaseAngle

increaseSize

decreaseSize

Fig. 4.14 Shape changing basis operations for tetrahedron networks

The shape changing basis operations generate or delete a tetrahedron or they change an angle and the size of a tetrahedron, respectively.

Chapter 4. Data Modelling and Management for 3D/4D Geoinformation Systems 81

addTetrahedron

takeOffTetrahedron

insertHole

deleteHole

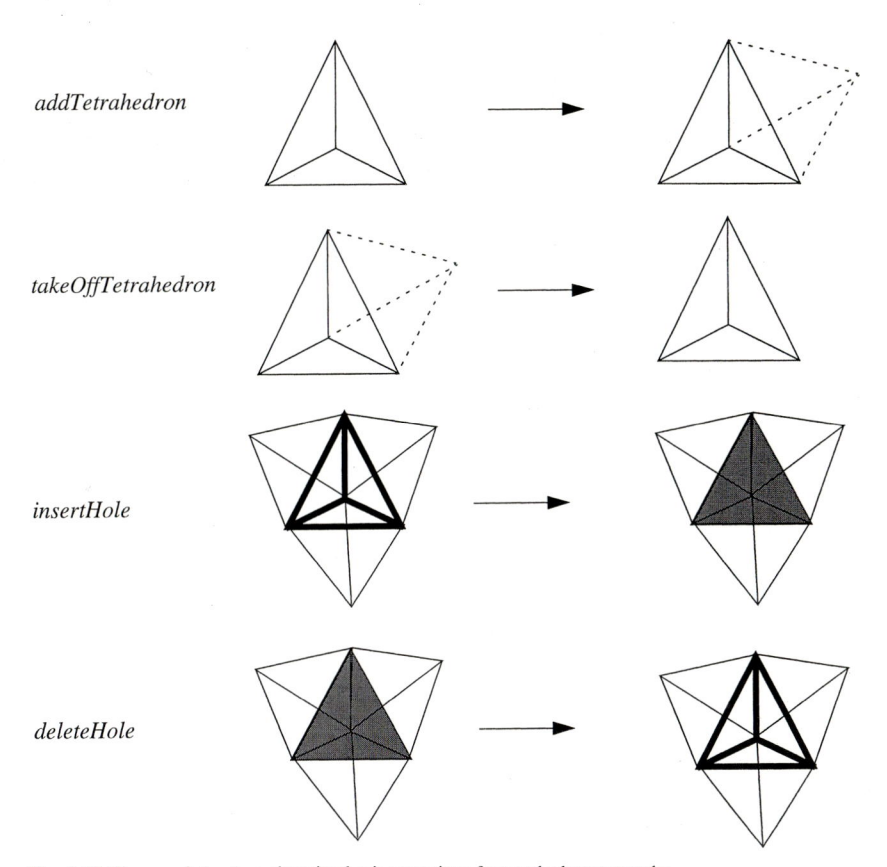

Fig. 4.15 Shape and structure changing basis operations for tetrahedron networks

Shape and structure changing basis operations cannot completely be separated from each other. Figure 4.15 shows basis operations that change the shape (geometry) as well as the interior structure (topology). This holds for the addition and deletion of a tetrahedron (*addTetrahedron* and *takeOffTetrahedron*) and for the insertion of a hole in the interior of four tetrahedra (*insertHole* and *deleteHole*). In Fig. 4.15 the front tetrahedron of the four tetrahedra has not been drawn, because it would have hidden the view to the other three.

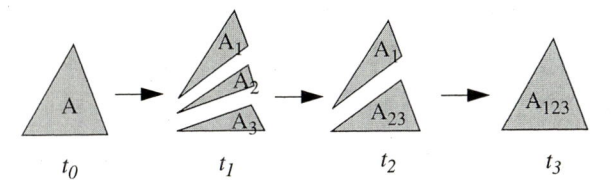

Fig. 4.16 Decomposition of a spatial object into single components and subsequent merging

Additionally, a spatial object can be decomposed into several single components and several single components can merge to a single object, respectively. In Fig. 4.16 we can see this change of the connected components. It shows that the change of the structure and of the shape cannot only be described by a pure geometry or topology change. Neither their definition points and their single coordinates nor the change of their discretizations describes the temporal sequence of the spatial object completely.

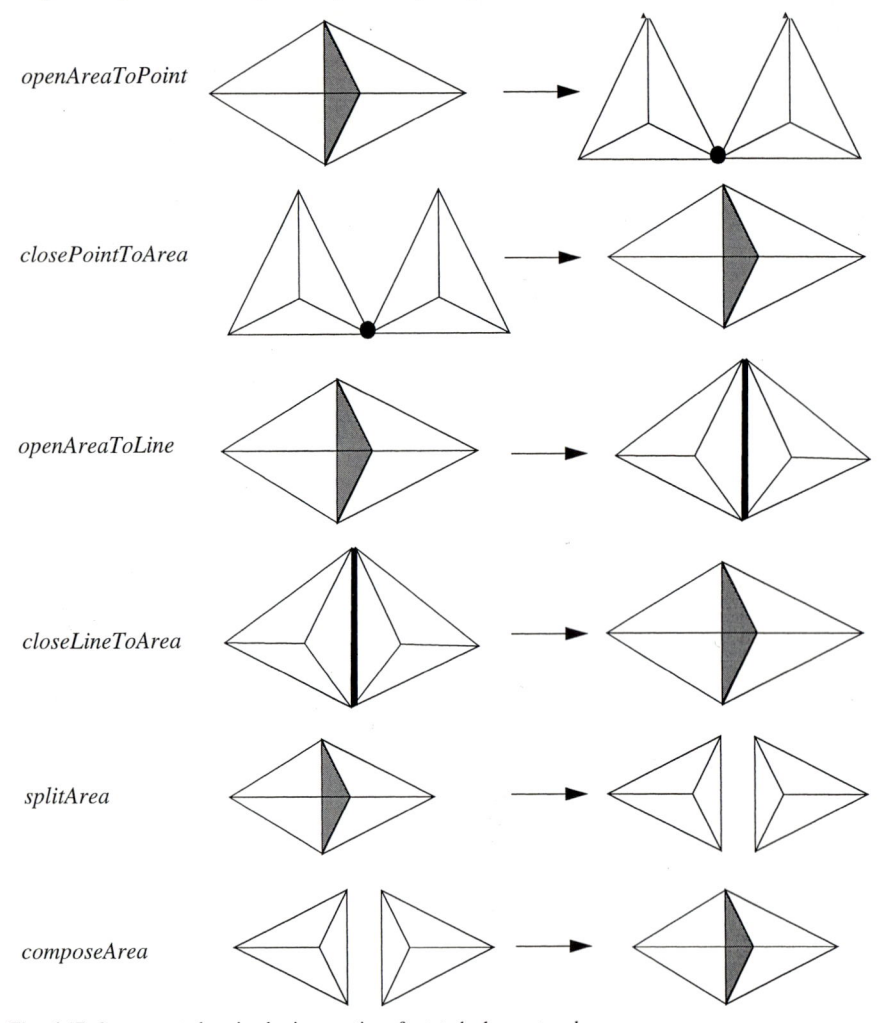

openAreaToPoint

closePointToArea

openAreaToLine

closeLineToArea

splitArea

composeArea

Fig. 4.17 Component changing basis operations for tetrahedron networks

Figure 4.17 shows basis operations that change the number of connected components of a tetrahedron network. We assume that for each component two tetrahedra must completely touch each other in an area (triangle); i.e. two tetrahedra that are only touching at a point or at an edge belong to two different components.

4.2.2.5 Temporal Topological Relationships

Besides the temporal development of the spatial topology we can also consider temporal topological relationships (see also the "travel time topology" of chapter 2.1.1.3). Figure 4.18 shows that the minimal set of binary topological relationships for two line objects A and B can directly be applied to the set of one-dimensional temporal relationships between two time intervals t_A and t_B. This observation is not new, such predicates for temporal intervals have already been examined by ALLEN (1983, 1984).

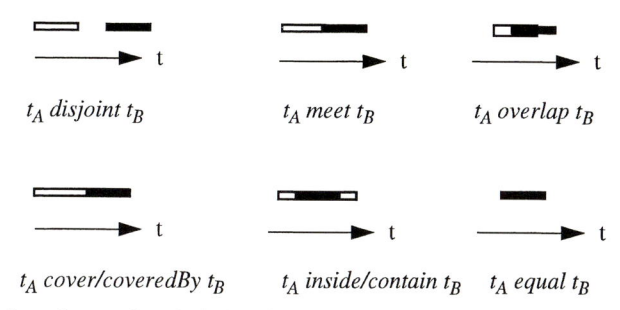

t_A disjoint t_B $\qquad$ t_A meet t_B $\qquad$ t_A overlap t_B

t_A cover/coveredBy t_B $\qquad$ t_A inside/contain t_B $\qquad$ t_A equal t_B

Fig. 4.18 Minimal set of temporal topological relationships between two time intervals t_A and t_B

The temporal topology describes the relative temporal relationship between two objects. The time intervals can be *disjoint*, overlap each other in different ways *(overlap, covers, inside)* or they can be identical *(equal)*.

4.2.3 Geo-Objects in Space and Time

The separated examination of geometry and topology in space and time from the thematic information is useful for the development of a space-time theory for geoinformation systems (PIGOT 1992b; HIRTLE and FRANK 1997; ERWIG et al. 1997a). However, it leads away from the observation of geoscientific phenomena and the modelling of processes as they are described in (SIEHL 1993; SPACCAPIETRA et al. 1998). Therefore we suggest a common modelling of thematic, topological and geometric information in spatio-temporal geo-objects.

We distinguish two ways to model the temporal dimension in geo-objects:

1. Time per attribute of a class, i.e. different attributes of a class can be time dependent. Each of these attributes has a set of states that are time dependent.
2. Time per object, i.e. each object of a class has a set of states that are time dependent.

For example, the attribute "geometry" can be seen as a function of time. Or the whole object is considered to be time dependent, i.e. spatial and non-spatial attributes are modelled as states of a geoscientific process.

It seems to be very useful for geoscientific applications to consider space and time as a unit in which geometry and topology can be observed as time dependent besides the thematic information.

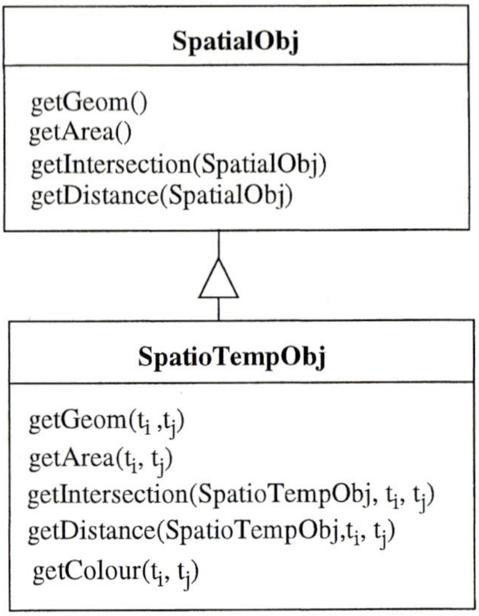

```
        ┌─────────────────────────────────────┐
        │             SpatialObj              │
        ├─────────────────────────────────────┤
        │  getGeom()                          │
        │  getArea()                          │
        │  getIntersection(SpatialObj)        │
        │  getDistance(SpatialObj)            │
        └─────────────────────────────────────┘
                          △
                          │
        ┌─────────────────────────────────────┐
        │           SpatioTempObj             │
        ├─────────────────────────────────────┤
        │  getGeom(t_i, t_j)                  │
        │  getArea(t_i, t_j)                  │
        │  getIntersection(SpatioTempObj, t_i, t_j) │
        │  getDistance(SpatioTempObj, t_i, t_j)│
        │  getColour(t_i, t_j)                │
        └─────────────────────────────────────┘
```

Fig. 4.19 Example of a class "SpatioTempObj" which is derived from the class "SpatialObj"

In Fig. 4.19 time and thematic information are added to the class *SpatialObj*. It returns the pure geometry and topology information. The class *SpatioTempObj* enables the access to geometry and to geometric funtions for discrete times. For that the user has to access the database in which the states of the objects are stored for certain defined times. For example, the function *getGeom(t_i, t_j)* returns the geometry for the set of object states that are stored between the times t_i and t_j. The function *getArea(t_i,t_j)* returns the time dependent area of the object and the functions *getIntersection(SpatioTempObj, t_i, t_j)* and *getDistance(SpatioTempObj,t_i, t_j)* return the intersecting geometry of two spatio-temporal geo-objects and their distances, respectively. The function *getColour(t_i, t_j)* returns the colour of the object for the states of the objects that are known between the times t_i and t_j.

As we have seen in the example above the temporal change of a geo-object can often visually be represented by a spatial change. Examples are the change of the area of a city or the restoration of strata blocks in geology.

It is straight forward to define the states for each single point of a geometry. Simpler and more efficient at saving storage space, however, is the generalization of the single points and definition of one state for all points of the geometry together. We will pick up this in chapter 4.3 in the context of the efficient management of spatial and temporal objects.

4.2.3.1 Equality of Spatio-Temporal Objects

Let us describe what we mean by the equality of abstract spatial objects in $\Re^3$ for the elementary geometric classes *point, line, surface* and *volume*. After that we will go into the simplicial complexes and we will define the equality of time dependent spatial objects.

Two abstract spatial objects are equal, if their *spatial structures and their spatial behaviours* are equal. The spatial behaviours are equal, if both objects have the same spatial operations with identical functionality. In the following we determine the equality of spatial structures under consideration of the same absolute location.

1. *Equality of points:* the spatial structures of two points $P_1 = (x_1,y_1,z_1)$ and $P_2 = (x_2,y_2,z_2)$ in $\Re^3$ are equal, if their geometries, i.e. their coordinates $(x_1=x_2)$ and $(y_1=y_2)$ and $(z_1=z_2)$ are equal by pairs.

2. *Equality of line segments:* the spatial structures of two line segments in $\Re^3$ are equal, if their (respective infinite) point sets are equal.

3. *Equality of surfaces:* the spatial structures of two surfaces in $\Re^3$ are equal, if their (respective infinite) point sets are equal.

4. *Equality of volumes:* the spatial structures of two volumes in $\Re^3$ are equal, if their hulls (boundaries) and their interiors are equal, i.e. if the (respective infinite) point sets of their boundaries and interiors are equal.

5. *Equality of polylines (approximated lines):*

the spatial structures of two polylines PL_1 and PL_2 in $\Re^3$ are equal, if their geometry, i.e. their base points, $(p_{11},...p_{1n})$ and $(p_{21},...p_{2n})$, as well as their topology, i.e. their connecting lines $((p_{11},p_{12}),...,(p_{1n-1},...p_{1n}))$ and $((p_{21},p_{22})...(p_{2n-1},p_{2n}))$ are equal by pairs.

6. *Equality of triangle networks (approximated surfaces):* the spatial structures of two triangle networks in $\Re^3$ are equal, if their geometries (base points) and their topology (triangle meshes) are equal and if the orientations of the normal vectors of (at least one of) their triangles are equal by pairs, respectively. We require that the triangulation is minimal, i.e. a coarser representation leads to another surface (geometry).

7. *Equality of tetrahedron networks (approximated volumes):* the spatial structures of two tetrahedron networks in $\Re^3$ are equal, if their geometry (base points), their topology (triangle meshes of their surfaces and the interior tetrahedron meshes) are equal. Notice that for objects with holes all the boundaries, the exterior and the interior ones, have to be considered. We require that the tetrahedralization is already minimal, i.e. a coarser tesselation leads to a change of the volume (geometry).

8. *Equality of spatio-temporal objects:* two spatio-temporal objects, i.e. two objects in $\Re^3$ which are variable in time are equal in a given time interval (t_0, t_n), if their spatial structures and their spatial behaviours are equal for all defined times $t_0, ..., t_n$. Especially geometry, topology and orientation of both objects must be equal during the whole time interval.

4.3 Management of Spatial and Temporal Objects

4.3.1 Checking of Spatial and Temporal Integrity Constraints

In geoscientific applications the checking of the integrity for objects in space and time is of central interest. By a *spatial integrity constraint* for a geo-object we mean an integrity constraint that automatically has to be executed by the database and that checks one or several of the following spatial aspects:

1. The absolute position of a geo-object and its extension (geometry);
2. The relative position of a geo-object to other geo-objects (exterior or global topology/metrics) or its internal spatial structure (interior or local topology);
3. The membership of a geometry or topology to a geo-object (existence).

In principle, the spatial integrity constraint is independent of any application. However, the constraint is mostly embedded into the context of a spatial application. For example, it can be checked if two geo-objects are intersecting each other or not. The decision as to whether the intersection is allowed in the application or not is the province of the application expert. It depends on the object types that are involved. For example, the intersection of two strata in geology is not valid, because physically only one stratum can be at the same place.

Making no claim to be exhaustive, a *temporal integrity constraint* for a geo-object is an integrity constraint that checks one or several of the following temporal aspects:

1. The absolute position of a geo-object and its spatial extension during the known states of a (periodically recurrent) time interval (geometry);
2. The relative position of a geo-object to other geo-objects (exterior or global topology/metrics) or its internal spatial structure (interior or local topology) during the known states of a (periodically recurrent) time interval;
3. The membership of a geo-object during the known states of a (periodically recuent) time interval (existence);
4. The time during which a geo-object exists in the known states of a (periodically recurrent) time interval;
5. The speed of a geo-object between two states.

If we reduce the time interval to a single state in the *temporal* integrity constraints (1) - (3), we get the *spatial* integrity constraints (1) - (3) which have been introduced before.

4.3.1.1 Checking of the Geometry

The geometry of a geo-object describes its absolute position and its extension in space and time. The absolute position in space can be controlled as follows:

1. Computation of the gravity point or the minimal bounding box of a geo-object and checking of its coordinates, for example inside an allowed three-dimensional region.

The extension of a geo-object in space can effectively be checked with the following integrity constraints:

1. "The extension (length, perimeter, area, volume) must not be zero and it must not exceed the value v";
2. "The maximal extension of the bounding box in x-, y- and z-direction must be inside the given interval (I_x, I_y, I_z)";
3. "Check if the geometry is convex or not".

The extension of the geometry for a geo-object in time, i.e. its geometry at a certain time or in the known states of a time interval can be controlled by the following additional integrity constraints:

1. "The geo-object cannot have existed longer than 100.000 years";
2. "In which time interval did the geo-object leave the allowed position?";
3. "In which time interval has the allowed extension of the geo-object fallen short of and been exeeded, respectively?".

4.3.1.2 Checking of the Topology

The topology of a geo-object describes its relative position to other geo-objects ("global topology") and its internal spatial structure ("local topology", for example the connection of points and edges of a triangulation) in space and time.

The relative position of a geo-object to other geo-objects in space can be controlled as follows:

1. Checking of the spatial position of a geo-object to other geo-objects with the minimal set of topological relationships *disjoint, meet, intersect, covers/coveredBy, inside/contain, equal* (EGENHOFER 1989).
2. Checking if the local topology of the boundary of a geo-object A "fits together" with that of a neighboured geo-object B. The hulls of the two geo-objects must exactly fit together, i.e. the triangles of both hulls must not intersect each other at any location. They must be identical in the intersecting region.

An adaptation of the geometries, e.g. with the discrete interpolation method (MALLET 1998b) is necessary, if the checking of the hulls of two neighboured geo-objects leads to a negative result.

Analogous to the spatial case, the relative *temporal* position of a geo-object to other geo-objects can be controlled with one-dimensional time intervals. The temporal integrity constraints check the binary topological relationships *(disjoint, meet, intersect, covers/coveredBy, inside/contain, equal)* for one-dimensional time intervals. It has to be checked, if the time interval in which a geo-object existed, is disjoint with the time interval of other geo-objects, if it intersects the time interval of other geo-objects, or if a time interval is inside another time interval. In the last case we have to distinguish, if both of the time intervals also start or end at the same time. Finally we have to check if both time intervals are identical.

It is also interesting to observe the *spatial topology* between geo-objects for different states *in time* to check the *integrity of the movements* for geo-objects.

Typical temporal integrity constraints are:

1. "If two geo-objects are *disjoint* at a starting state S_0 and if they *overlap* each other at a final state S_n, then there must have been an intermediate state S_i ($0 < i < n$) in which the two geo-objects touched each other from outside *(meet)*".

2. "If two geo-objects are *disjoint* at a starting state S_0 and if they intersect each other at a final state S_n for the first time *(overlap)*, then they must not have been equal to any state S_i ($0 < i < n$)".

3. "If a geo-object has been completely inside another at a starting state S_0, then at the next state transition only one case can occur: the first object is touched by the second from outside *(coveredBy)*".

In Fig. 4.10 and Fig. 4.11. we have already presented the spatial state transitions occurring bet-ween topological relationships of moving objects.

4.3.2 Spatial and Temporal Database Queries

Making no claim to develop our own spatio-temporal extension for a geodatabase query language, we intend to show how spatio-temporal "building blocks" can be integrated into existing query languages.

One of the most used types of GIS functions are ad hoc queries for thematic, spatial and temporal data. It is often useful to visualize the query formulation (input of the query) and the query result (output of the query). Extensions of geometric data types for spatial query languages have been proposed by (GÜTING 1989; SVENSSON and ZHEXUE 1991; GÜTING and SCHNEIDER 1993; EGENHOFER 1994) and by other authors. For example, some extensions for temporal predicates are known as TQUEL (SNODGRASS 1987) and TSQL2 (SNODGRASS 1995). T/OOGQL (VOIGTMANN et al. 1996) is a prototype of a spatio-temporal query language for GIS.

We extend the approach of (BREUNIG 1996) for temporal database queries and show that spatial as well as temporal database queries can be decomposed into typical "building blocks". We introduce the spatio-temporal attributes, spatio-temporal predicates, spatio-temporal functions and spatio-temporal operators as the elementary building blocks. Examples for spatio-temporal attributes are the volume of a solid and the time interval in which it existed, respectively. Examples for spatio-temporal predicates, spatio-temporal functions and spatio-temporal operators are given below. The result value of a spatio-temporal predicate is of type boolean. A spatio-temporal function returns an integer or a real value and a spatio-temporal operator returns a geo-object that can be used recursively as an argument in other spatio-temporal operators. Following the Relational Algebra (CODD 1970) and extending

Chapter 4. Data Modelling and Management for 3D/4D Geoinformation Systems

the approach of GÜTING (1994), we introduce the following further "building blocks": the spatio-temporal selection, the spatio-temporal function application and the spatio-temporal join.

A spatio-temporal selection selects those geo-objects of an object set that has been specified by one or more spatio-temporal predicates. The spatio-temporal function application returns the function value of a spatio-temporal function. Finally, a spatio-temporal join is the comparison of a spatio-temporal attribute which identically has to be defined in two object sets, respectively. If the predicate is true, the new "combined object" is qualified for the resulting object set.

We show the combination of building blocks for spatio-temporal database queries with respect to three examples in natural language and in a SQL-similar notation.

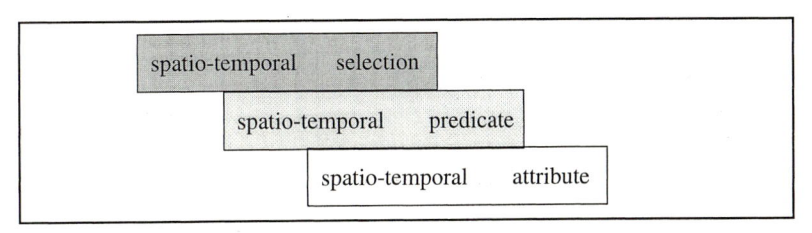

Fig. 4.20 Combination of spatio-temporal selection, predicate and attribute

The following three examples refer to the combination of building blocks in Fig. 4.20.

Example 1:

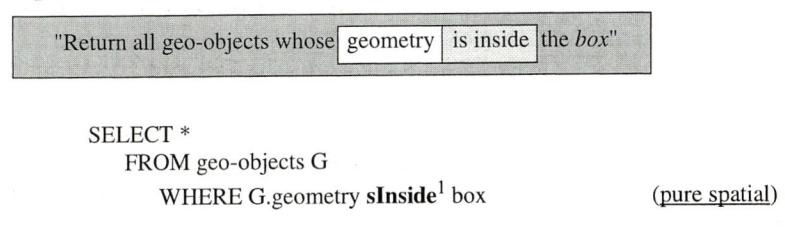

"Return all geo-objects whose geometry is inside the *box*"

        SELECT *
           FROM geo-objects G
              WHERE G.geometry **sInside**[1] box   (pure spatial)

Example 2:

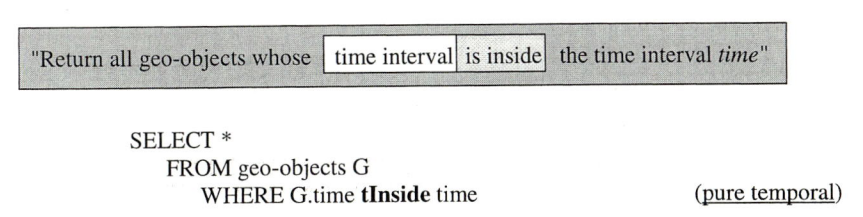

"Return all geo-objects whose time interval is inside the time interval *time*"

        SELECT *
           FROM geo-objects G
              WHERE G.time **tInside** time   (pure temporal)

---

[1] The prefix "s" (sInside etc.) stands for "spatial", the prefix "t" in the following temporal predicates (tInside etc.) for "temporal".

Example 3:

"Return all geo-objects whose geometry is inside the *box* during the time interval time"

> SELECT *
> FROM geo-objects G
>   WHERE (G.time **tInside** time)
>   AND (G.geometry **sInside** box)   (spatio-temporal)

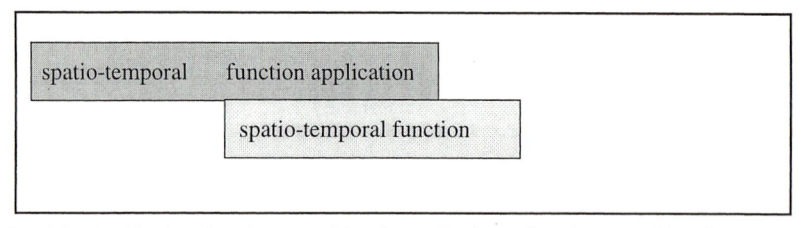

spatio-temporal function application

spatio-temporal function

Fig. 4.21 Combination of spatio-temporal function application and spatio-temporal function

The following examples refer to the combination of the building blocks in Fig. 4.21.

Example 1:

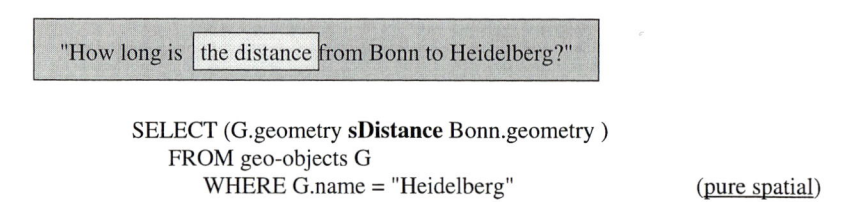

"How long is the distance from Bonn to Heidelberg?"

> SELECT (G.geometry **sDistance** Bonn.geometry)
> FROM geo-objects G
>   WHERE G.name = "Heidelberg"   (pure spatial)

Example 2:

"How long is the needed time distance (by train) from Bonn to Heidelberg?"

> SELECT (G.geometry **tDistanceTrain** Bonn.geometry)
> FROM geo-objects G
>   WHERE G.name = "Heidelberg"   (pure temporal)

> "How long is the distance from Bonn to Heidelberg and how long is the needed time distance ?"

UNION (
        SELECT (G.geometry **sDistance** Bonn.geometry)
           FROM geo-objects G
              WHERE G.name = "Heidelberg",
        SELECT (G.geometry **tDistanceTrain** Bonn.geometry)
           FROM geo-objects G
              WHERE G.name = "Heidelberg"   (spatio-temporal)
)

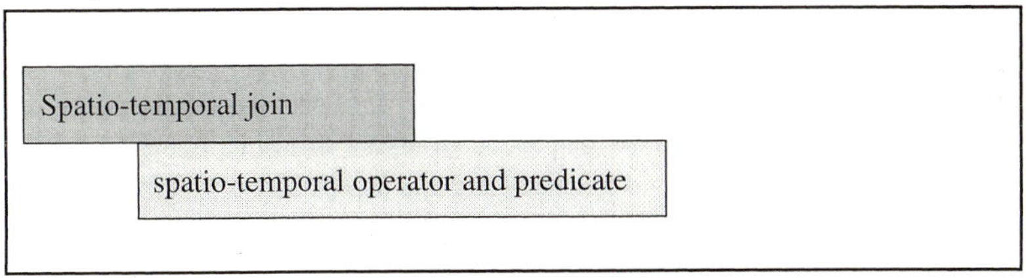

> Spatio-temporal join
>   spatio-temporal operator and predicate

Fig. 4.22 Combination of spatio-temporal join, operator and predicate

The following three examples refer to the combination of the building blocks in Fig. 4.22.

Example 1:

> "Return the recent intersections of geological strata and faults in the Lower Rhine Basin, i.e. all stratum-fault pairs that are intersecting ".

        SELECT (Stratum.geometry **sIntersection** Fault.geometry)
           FROM geo-objects Stratum, geo-objects Fault
              WHERE Stratum.geometry **sIntersect** Fault.geometry   (pure spatial)

Example 2:

> "Return the temporal intersection of geological strata and faults, i.e. all stratum-fault pairs that are temporally overlapping each other"

        SELECT (Stratum.time **tIntersection** Fault.time)
           FROM geo-objects Stratum, geo-objects Fault
              WHERE (Stratum.time **tIntersect** Fault.time   (pure temporal)

Example 3:

> "Return the stratum-fault pairs with their overlapping time intervals that today are spatially intersecting each other"

```
SELECT (Stratum.time tIntersection Fault.time)
FROM geo-objects Stratum, geo-objects Fault
WHERE (Stratum.geometry sIntersect Fault.geometry
```
(spatio-temporal)

The decomposition into spatial and temporal building blocks shows well how spatial and temporal predicates, functions and operators are embedded into database queries. The spatial and temporal selection, the function application and the join serve as a framework. In this framework the optimization of spatial and temporal database queries can be executed according to the rules of algebraic optimization (e.g. CERI and GOTTLOB 1985). In this technique SQL expressions are decomposed into smaller units. However, additionally the complexity of the embedded geometric and temporal functions and operators has to be considered. Their complexity can even invert the execution order of selection, function application and join, respectively. We will later enter into the optimization of 3D operators in the context of a geological application (see chapter 5.4.3). First we examine the problem of optimizing the execution of spatial and temporal predicates. Therefore a geodatabase needs spatial and temporal access methods.

4.3.3 Spatio-Temporal Database Access

Following the notions in chapter 2.1.1 we generally distinguish between the field-based and the object-based spatial and temporal database access. In the field-based approach space is decomposed into regular cells (SAMET 1990; NIEVERGELT and WIDMAYER 1997) or time is decomposed into regular time intervals. The objects are distributed across these time intervals. An example of a field-based spatial access method is the quadtree (FINKEL and BENTLEY 1974; SAMET 1990) or in three-dimensional space its extension to the octree.

Against that, in the object-based or data driven approach the structure of the space and time partition is dependent on the insertion order of the objects into the access method. The size of the cells is increasing if the objects are extending and they have to be split, if the maximal size -which has to be determined à priori- is exceeded. Examples for object-based access methods are R-Trees (GUTTMAN 1984; SELLIS et al. 1987; BECKMANN et al. 1990). GAEDE and GÜNTHER (1998) give a good overview of existing spatial access methods.

The Hextree (HAZELTON et al. 1990) is a simple spatio-temporal access method which is based upon the field-based approach. Following the octree approach it uses four-dimensional fields for an efficient spatio-temporal access. The Hextree can easily be realized on top of a B*-Tree. However, this access method has obvious disadvantages concerning the access for objects with a large spatial or temporal extension. They have to be stored at the topmost level of the Hextree. This can lead to a fast overflow of the Hextree nodes which can result in an overflow of the associated database pages.

In the following we refer to the object-based access approach. Concerning the spatial access we assume that the problem is to search a spatially specified small set of objects within a large object set. We also assume that each object has stored a minimal circumscribing box (MCB) as its approximation or that the MCB can easily be computed with the geometry of the object. The spatial query predicate usually is a box whose surfaces are parallel to the x-, y- and z-axes. These simple geometries of the query predicate and of the stored objects lead to an efficient spatial access. The only operation needed is the intersection or containment of two boxes.

The algorithms for the management of two-dimensional object-based access methods like the R*-Tree (BECKMANN et al. 1990) can directly be extended to three-dimensional space. The following extensions are necessary:

1. Extension of the minimal circumscribing rectangles (MCR) to minimal circumscribing boxes (MCB);
2. Extension of the two-dimensional query window to a three-dimensional query box;
3. Consideration of (1) and (2) for the geometric comparisons between the MCBs and the query box in the management algorithms *insert, delete* and *retrieve;*
4. Consideration of (1) and (2) to determine a suitable measure for the minimal space used and the overlap between the cells (*insert* and *split*).

Let us give an example. To realize an R*-Tree (BECKMANN et al. 1990) in three-dimensional space, the three-dimensional measures of the minimal volume, the minimal surface and the minimally overlapping volume have to be taken instead of the corresponding two-dimensional measures in point (4).

However, there are some problems with the transition from two- to three-dimensional space for the management of low-dimensional objects, i.e. of one- or two-dimensional objects in three-dimensional space. If a line or surface object is parallel with two of the three axes of the coordinate system, then the volume of its MCB is zero. Thus it cannot reasonably be inserted into a node of the spatial access method. For such "flat MCBs" we introduce a special treatment which considers the original 2D measures instead of the 3D measures. We use the minimal area and the minimal perimeter of each object as well as the minimal overlapping area of the objects in one cell.

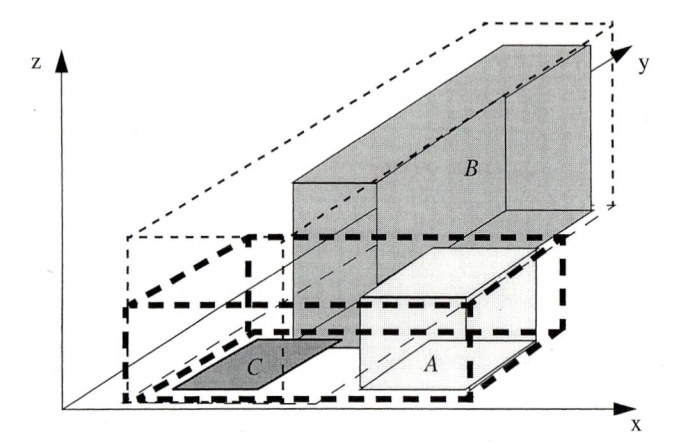

Fig. 4.23 Special treatment for the insertion into the R*-Tree of a plane area C which is parallel to the x-y plane

Figure 4.23 shows an example in which the MCB of object C is parallel to the x-y plane. In the split algorithm, for example, it has to be decided if the MBC of object C should be inserted into the cell of object A or B. In the shown case the MBC of A need not be increased as much as that of B to insert A. Furthermore, the overlapping volume is smaller if A and C are put together than is the case with A and B. That is why the objects A and C are put together into one cell. The corresponding box is drawn bold dotted in Fig. 4.23. The alternative, to store C together with B is drawn finely dotted in the same figure.

The *spatial clustering* of the objects, i.e. the approach to store the objects on disk in such a way that their spatial neighbourhood is obtained -if possible- is an important aspect for the implementation of spatial access methods.

It can also be very useful to cluster whole query paths for the support of often used database queries. This means storing several nodes of the tree together on one page or cluster in secondary storage. The overlap of the MCBs has to be minimized. This is extremely difficult if objects with variable size have to be managed. Figure 4.24 shows the clustering of triangles into two groups after a split in the R*-Tree and the R-Tree, respectively. There is no solution in this example for which the entries in the nodes are equally distributed without an overlap of the MCBs[1]. The solution at the right side (R-Tree) obviously shows a larger overlap than the solution for the R*-Tree at the left side. The reason for that is that the R*-Tree considers the nearness of the objects by minimizing the overlapping areas.

---

[1]. The examples, however, are only shown in two-dimensional space to simplify the presentation.

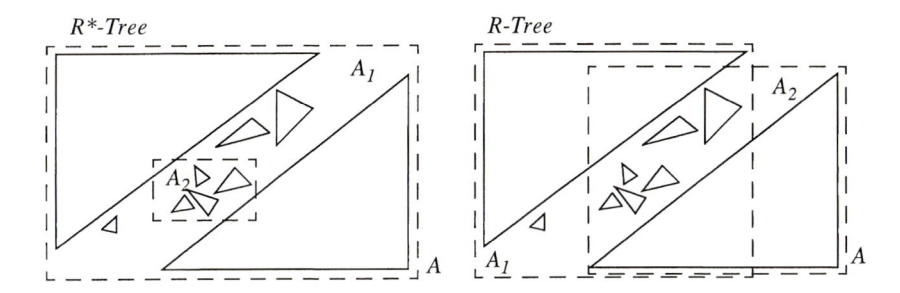

Fig. 4.24 Examples for the clustering after the split of a node A into two nodes A1 and A2 in the R*-Tree and the R-Tree, respectively

For the spatio-temporal data access we assume that a sequence of object states at different times t_i, $(i = 0, ..., n)$ is stored in the database. An example is the states of a moving stratum in a geological process. The simplest method to manage a temporal sequence of object states is to use a spatial index structure for each time and all objects that are stored at the respective time. An example of such a realization is multiple R-Trees (MR-Trees). They store a sequence of R-Trees at each time t_i with their roots organized in an array temporally ordered (XU et al. 1990). It is possible to access these search trees from several roots (shared subtree concept), if the position of an object has not changed within the corresponding time steps.

The MR-Tree supports the following types of spatio-temporal queries:

1. Queries for objects inside a specified spatial region (without time information);
2. Queries for objects inside a specified temporal interval (without space information);
3. Queries for objects that have been inside a specified region within a specified time interval (*intersect* and *contains*);
4. Queries for objects that have changed their location from a time t_i to a time t_j or for objects that stayed at the same location.

The disadvantage of an MR-Tree, however, is that the management costs are too high because of the use of several R-Trees in parallel. For pure temporal queries in most cases several R-Trees have to be used to answer the queries.

Obviously it is better to integrate the time intervals of the objects besides the MCBs in the key of each node. Then only one combined spatio-temporal index tree has to be used. In the root of the tree then the largest temporal interval ("universe") is stored. It has to be dynamically updated, if the temporal change of the objects in the leaf nodes and index nodes demands the update. The structure of such a spatio-temporal access method, however, should be influenced by the spatial extension of the inserted objects. An exception should be made, if the temporal access plays a significant role in a given application. In this case the R-Tree can be extended for the temporal access.

The RT-Tree (XU et al. 1990), a height balanced tree, is an example for such a solution. The entries of its nodes consist of a triple *(S, T, P)* with *S* as the spatial key, *T* as the temporal key and *P* as the pointer to the corresponding sub-tree index nodes or as the pointer to the physically stored objects in leaf nodes.

Depending on the preference of the spatial, temporal or thematic database queries, for the RT-Tree different split strategies have been proposed. They split a node according to the spatial or temporal extension or according to additional knowledge about spatial or temporal relationships. Examples are a split along a known spatial or temporal boundary like a border between two countries or the temporal border between 20th and 21st century.

We can speed up the access to temporal intervals, if we orientate the split strategy towards the temporal intervals. However, we then lose efficiency for the spatial access. It is a particular problem to support queries that consider large time intervals like "Return all objects that have been inside the given box in the complete temporal universe (x_0, x_n)". That is why we propose for applications that mostly require temporal queries or spatio-temporal queries with large time intervals to use a one-dimensional index like the B-Tree (BAYER and MC CREIGHT 1972). Then the spatial and the temporal filter can be used independently of each other. The spatial position is then considered separately[1]. This is particularly useful for queries to the temporal sequence of states of a single object. Queries of this type are only inefficiently supported by mixed spatio-temporal access methods. In the worst case the whole tree has to be searched through. An example is an object that spans the whole spatial universe to each time that is stored in the database.

The RT-Tree is well suited then, if the spatial component of queries is more important than the temporal access. For the temporal predicate, however, eventually large parts of the tree have to be searched through.

As a problem we have recognized that objects which have been spatial neighbours at the beginning of the observation time can have a large distance between them at the end of the observation time. Thus there is no "optimal 4D access method". In the general case the approach of extending a three-dimensional access method "by the time dimension" does not lead to success. Further approaches that effiiciently support the spatio-temporal access with R-Trees are the 3D-Tree (THEODORIDIS et al. 1996) and the HR-Tree (NASCIMENTO and SILVA 1998). Both extensions, however, are reduced to the management of moving point data in two-dimensional space and also they do not solve the problem of large time intervals.

As the data volume is quickly increasing in many time-dependent applications, the minimal bounding box (MBB) provides only a poor approximation for the representation of the geometries in spatio-temporal indexes. With this coarse filter too many "false hits" occur for queries with large spatial and temporal extensions. SCHNEIDER and KRIEGEL (1991) and LEE et al. (1997) have already proposed tesselation techniques for complex objects. However, they are reduced to two-dimensional space. We extend the recursive tesselation technique of LEE et al. (1997) as follows for four dimensions, i.e. three spatial and one temporal dimension.

---

[1]. The precondition for this approach is that all objects are using the same time.

Chapter 4. Data Modelling and Management for 3D/4D Geoinformation Systems 97

We first show the technique with the tesselation of closed polygons in two-dimensional space (Fig. 4.25).

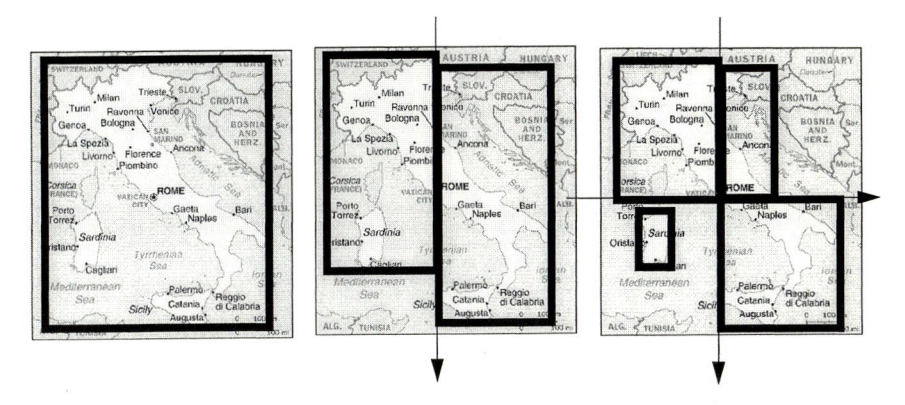

Fig. 4.25 Tesselation technique of LEE et al. (1997) shown with the example of a map from Italy[1]

The region consisting of Sardinia and the main part of Italy has been subdivided alternately, each time vertically or horizontally in the middle of the region. The tesselation of the minimal bounding box (MBB) in each direction (vertically or horizontally) terminates if the new MBB has less or equal 2^{-g} of the area. Let g be a parameter that has a priori been determined for the maximal tesselation depth. Taking $g = 2$, for example, the method terminates if the corresponding MBB contains less than 25% of the whole space. Depending on the subdivision, new MBB are computed for the new components. In the example of Fig. 4.25 the subdivision terminates after the second step, because particularly the components left below and right at the top contain less than 25% of the whole area. The single MBB can be managed with a binary tree. Each node contains the left and the right successor node (index node) and the pointer to the MBB of an object (leaf node), respectively.

It is obvious that the storage space saved with the tesselation of the geometries is even higher for objects in three-dimensional space than in the plane, especially for surfaces in three-dimensional space. Space is then alternately divided in half, parallel to the x-, y- or z-axis. The new components again can be managed with a binary tree. For example, the number of components to be managed by the tree are 2^3 after three tesselation steps.

In a similar way we can execute a tesselation along the axes for moving objects in 3D space. Especially for large time intervals, i.e. for objects that exist for a long time, the tesselation leads to a significant reduction for the number of false hits in time-dependent database queries refering to a small spatial query box.

---

[1]. The map has been taken from the public map archive http://www.lib.utexas.edu/Libs/PCL/Map collection/Map collection.html

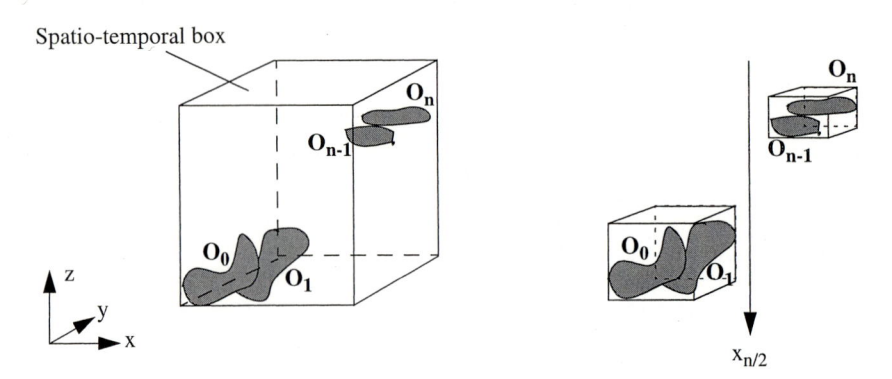

Fig. 4.26 Tesselation of moving objects in 3D space

In Fig. 4.26 we assume that the state of an object is stored in the database for the times t_0, t_1 and t_{n-1}, t_n. Then for a time-dependent database query with the time interval (t_0, t_n) the MBB of the object have to be computed for all times. With the tesselation along the middle of the x-axis $(x_{n/2})$ into two MBB the empty space is significantly reduced. Those times can be omitted for which no objects are stored. Thereby fewer paths have to be searched through in the R-Tree.

The result is that we can reduce the search space by applying the tesselation technique. However, for applications in which the objects are moving large distances, we have to decide if the spatial or the temporal access should efficiently be supported. If we particularly support the spatial access, then the temporal order of the nodes in the tree is considered only in a subordinate way. To achieve an efficient temporal access, however, in principle the spatial neighbourhood of the objects cannot be considered. Furthermore, the data volume is quickly increasing, if for each point of a geometric object its own state is attached at every time. Thus it is useful to refer the state to the whole object or its approximation, respectively. Even with this realization we have to be careful with large objects (e.g. 1,000,000 triangles or more), because the data should not be stored redundantly (i.e. the complete data in each state): but, if possible, only the updates from some last states should be stored with the new state. The update can either be computed by the direct predecessor version (delta storage) or we can refer to a stored "full version" some times before. Such full versions can be stored at fixed times. An object is subdivided into sub-objects, which have been changed at certain time intervals. Such a spatial partition for temporally changeable objects is an alternative to compression techniques, for example of triangle networks. It is particularly useful if the size of one of the objects already exceeds the size of main memory. In this case it is necessary to access parts of the object after the unzipping of the objects.

4.3.4 Supporting the Visualization for Large Sets of Spatio-Temporal Objects

The animation of geo-objects requires a large data set that cannot permanently be kept in main memory. Thus secondary storage methods being oriented for spatial and temporal storage should also be used in geoinformation systems to support efficiently the visualization of the objects. The requirements for spatial and temporal access methods particularly refer to:

1. The loading of parts from large objects;
2. The loading of an object set that has to be visualized (fly through) in a scene;
3. The consideration of differently resolved objects, i.e. coarsely resolved in the background and finely resolved in the front;
4. The consideration of aspect (3) depending on the reference point and the view angle of the observor.

The aspects (1) and (2) will even become more important for future geoinformation systems, because the data sets to be visualized will further increase. In spite of increasing main memory sizes, the data cannot all be kept in main memory during data visualization. Aspect (1) is important to support the visualization of very large objects (e.g. > 10,000 triangles), if only a small part of them has to be analysed in detail. The spatial access method quickly has to determine those parts of the object that are inside a predetermined three-dimensional region or that intersect that region. For each loading of such a region only one query is necessary in an R*-Tree.

More costly to realize is aspect (3) which requires the support of *perspective region queries* by the spatial access method. It is important that we can resolve more coarsely those objects that are further away from "fly over" scenes. We can show them in detail if they are close to the location of the observer. The resolution of the objects should also be coupled to the view angle of the observer. This means that for the access method "perspective region queries" have to be supported and the result has to be passed directly to the visualization. The result of a perspective region query are all objects that are inside the minimal bounding box of the chosen view cone. Figure 4.27 shows the principle of a perspective region query.

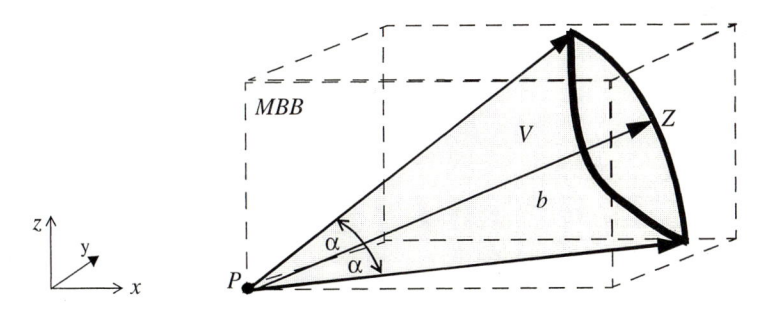

Fig. 4.27 Principle of a perspective region query

There is information that goes into the region query which usually is only used during the visualization of the objects: starting from the position P of the observer a view cone V with opening angle α is spanned, which ends at the goal point Z. The observer focuses to Z. The view vector b is computed with $b = z\text{-}p$. The minimal bounding box of the view cone serves as the query box; i.e. it determines the region that is relevant for the perspective region query.

It is a problem to keep the representation of the result set small in main memory for large Z or large α. Therefore we must resolve the objects with a large distance from the observer only coarsely, whereas the near objects have to be represented in detail. In the following we use three user-defined abstraction levels: one for the original representation, one for a middle approximation and one for a coarse approximationn of the objects. For each object we use its own R*-Tree per abstraction level[1].

The R*-Tree distributes the result of the perspective region query to the three different distance zones that correspond to the three degrees of detail. The near objects are returned by the spatial acess method as finely approximated objects. The objects with a middle distance are returned as objects with a middle resolution and the objects which are a large distance from the observer are returned and visualized as coarsely approximated objects. Figure 4.28 shows the principle of such a perspective region query which is extended to different distance zones. Depending on the view direction and the view opening angle the zones are intersecting each other differently.

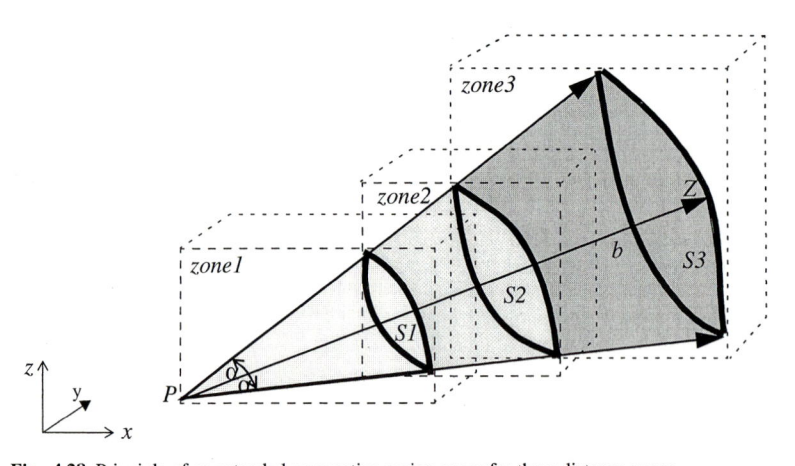

Fig. 4.28 Principle of an extended perspective region query for three distance zones

Starting from the position and the view direction of the observer the zones can be determined as follows: *zone 1* is determined by the MBB of view cone 1, *zone 2* by the MBB of the difference between view cone 2 and view cone 1, and *zone 3* is determined by the MBB of the difference between view cone 3 and view cone 2. Figure 5.9 in the next chapter shows an example of a perspective region query for geological data in the Lower Rhine Basin. The three different distance zones can clearly be seen at the different resolutions of the triangle networks.

The user has to decide in advance for objects that are inside the overlapping regions, if the finer or the coarser degree of detail has to be selected. The other degree of detail that has not been selected has to be locked in the R*-Tree.

---

[1]. In the GeoToolKit implementation of the "Multi level R*-Tree" (WAGNER 1998) the corresponding approximations are fixed with 2%, 10% and 100% as the degrees of detail for the elements (triangles) of the original data.

The retrieve function of the extended perspective region query returns a set of pointers to the objects that are inside the view cone of the objects. The retrieve function gets as input parameters the point P of the observer, the goal point Z, the opening angle α and the three distances *dist1*, *dist2* and *dist3* for the distances between the position of the observer and the end of the nearest, the middle and the longest distance zone.

The same mechanism of the region query that is controlled by different distance zones could also be used for objects that are moving in time. For each time t_i in which objects are stored in the database, the distance of the objects to the observer has to be computed. That is how the relationship of the objects to the corresponding distance zones of the R*-Tree has to be determined.

Chapter 5

Systems Development: from Geodatabase Kernel Systems to Component-Based 3D/4D Geoinformation Systems

Starting with a short introduction of the first geodatabase kernel systems, in this chapter we pick up the concepts of chapter 4 and show implementations with the so called GeoToolKit. Finally specific components on top of GeoToolKit are discussed for a geological application in the Lower Rhine Basin, Germany.

5.1 Geodatabase Kernel Systems

5.1.1 Requirements

As we have already seen in the last chapters, geoscientific applications require from a database the support of spatio-temporal data types and access methods. They can be integrated into the data model, the query language and the physical storage manager of a geodatabase. Geo-applications need different and specific geometric data types. All the same, more than 10 years ago it was recognized that it is useful to identify a "kernel" of geometric standard data types (SCHEK and WATERFELD 1986; WOLF 1989; WATERFELD and BREUNIG 1992). They can be realized in a "geodatabase kernel" and are treated in the same way as other standard data types like integer, character etc. in the database management system.

Special requirements of geo-applications for a database kernel are:

1. From a database point of view, k-dimensional access methods (k > 1) are required to access spatial and temporal data. What is needed are efficient access methods for the (combined) spatial, temporal and thematic access. They have not been available in former standard DBMS.

2. From a software development point of view, implicated by the persistent nature of the data, the extensibility and flexibility are essential. This enables the incremental integration of new geometric classes into the standard class library without modification of already existing data and algorithms.

3. From a user point of view a flexible user-interface with a 3D/4D visualization component is required. It allows to integrate new modules into the existing implementation of the visualization. For example, it should easily be possible to integrate the visualization of objects represented with a boundary representation (B-Rep), if the visualization of volumes is already realized. Finally, it should be possible to formulate database queries to these new visualized objects.

From these requirements two new objectives can be derived: the first is that *the design and the implementation of geo-applications should be made easier*. Secondly, a geodatabase kernel should provide the *basis for an integration of different geo-applications*.

A geodatabase kernel should be general enough to be used in different application types. The generality and extensibility of its class library is important, so that it can easily be adapted to concrete applications. At the same time a geodatabase kernel should be "more than an empty interface specification". It should provide geometric operations and mechanisms to determine the position of selected objects in space and time.

5.1.2 The DASDBS Geokernel

5.1.2.1 Concepts

Support for the Management of "Complex Objects"

In the middle of the eighties database research has started to use the notion of "complex objects". In contrast to the "flat" tuples in Relational Databases they already contained hierarchically structured records. The notion of the object only referred to the structure, but not as yet to the methods of an object. DITTRICH (1986) called this aspect the "structural object orientation". The DASDBS[1] Geokernel (SCHEK and WATERFELD 1986; WOLF 1989; WATERFELD and BREUNIG 1992), one of the first database kernel prototypes for the support of geo-applications, has treated complex objects as tuples of nested relations, described with the so called NF^2-data model (JAESCHKE and SCHEK 1981; SCHEK and SCHOLL 1986). With the set-oriented interface (WEIKUM et al. 1987) to the DASDBS (PAUL et al. 1987) a set of complex objects could be stored by a single database call. Furthermore, large complex objects could be - as best possible- stored on a neighbouring database page. Thus joins between objects and sub-objects were "precomputed". The result of a database query then is a set of complex objects that can be provided for the geo-application in main memory by an object buffer.

Levels of Extensibility
1. Externally Defined Geometric Data Types

The DASDBS Geokernel facilitates the application programmer to integrate user-defined geometric operations into the database system. The operations then can be called by a query or by a spatial access method. The so called externally defined data types (EDTs) have to be made known by the geokernel. With this technique the DASDBS Geokernel can store objects of an EDT and retrieve them again (WOLF 1989). The EDT is treated in the same way as other standard data types like STRING or ORDINAL.

GEOM2D is an example of a class for two-dimensional geometries in the geokernel. It provides interfaces for geometric operations like *<edt>cli* (clip) and *<edt>cbb* (compute bounding box). The two operations mentioned, intersect a two-dimensional geometry at a window and compute the minimal circumscribing rectangle of a two-dimensional geometry, respectively. Within the three-dimensional geometry type GEOM3D the types G3PKT, G3LPG, G3FPG and G3FRS have been defined in the geokernel where they stand for three-dimensional points, polygon lines, oplygon areas and elevation rasters (only surfaces). However, three-dimensional applications have only partially been evaluated during the development of the DASDBS Geokernel (WATERFELD and SCHEK 1992).

---

[1]·DASDBS stands for *"DArmStadt DataBaseSystem"*

Chapter 5. Systems Development 105

2. Spatial Access Methods

The most used database access in geo-applications is the spatial access to a set of geo-objects. For this spatial access in the geokernel three different spatial access methods have been implemented: a quadtree based on B*-Tree (SPIES and SPEVACEK 1990), a Grid File (DOERPINGHAUS 1989) and an R-Tree (BREUNIG 1989). They have been realized as "direct" as well as "indirect" spatial access paths, i.e. they allow the spatial indexing as well as the *spatial clustering* of the geo-objects themselves. It is also possible to extend the geokernel by new access methods. At the interface to the so called access manager an index can be generated. The parameters needed are the transaction id, the relation id, the list with the attribute numbers of the attributes (which should serve as keys of the indexes) and the parameters needed for the index access method. The retrieval *(am_retrieve-tup)* then is executed set-oriented with a so called transfer area in which the result tuples are inserted. A performance comparison for the retrieval with the three access methods Quadtree, Grid File and R-Tree with data from a map of Germany has been carried through by WATERFELD (1991). It has been shown that the R-Tree is tendentially the best access method for large query windows and a high degree of data overlap. The Grid File is well suited for smaller query windows and a smaller amount of data overlap. Finally the quadtree is best suited for medium large query windows and for larger data overlap.

5.1.2.2 Architecture

The externally defined data types and spatial access methods of the geokernel are integrated in the so called access manager. Figure 5.1 shows the architecture of the geokernel and its embedding into DASDBS.

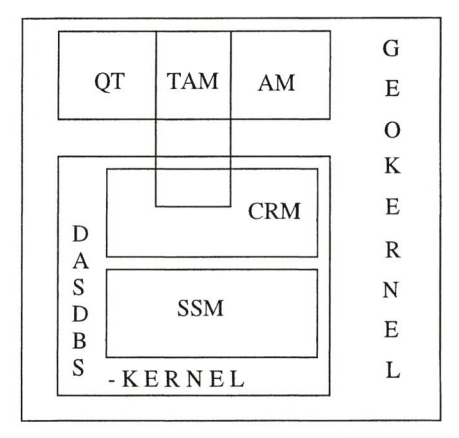

Fig. 5.1 Architecture of the geokernel and its embedding into DASDBS (from: BREUNIG et al. 1990)

The user-interface consists of the three modules of the access manager (AM), the transaction manager (TAM) and the query tree (QT). The AM contains the real interface of the geokernel. The TAM provides the managing operations of the object buffer. With their help tuples can be transported from main memory to the object buffer and the other way around. The storage region (transfer area) plays the role of a data type. A variable of the type "tfa" (transfer area) is handed over to the DASDBS kernel, comparable with the parameter transfer of an integer variable. The data structure for the query formulation has to be specified by a query tree.

5.1.2.3 Coupling the DASDBS Geokernel with a Map Construction System

The most advanced application of the DASDBS Geokernel has been realized by the coupling of the map construction system THEMAK2 (GRUGELKE 1986) with the geokernel (WATERFELD and BREUNIG 1990). This coupling has been implemented within the DFG priority programme "Digital Geoscientific Maps" (VINKEN 1988, 1992). It shows well how the data handling of the former file system in the GIS can be replaced by a geodatabase kernel.

Figure 5.2 shows the concrete architecture of THEMAK2. The data handling component and the graphical kernel system are directly based upon the file system. The essential module of the *user-interface* is the command language. In this module maps can be generated with single commands like the drawing of a line or the generation of windows. In THEMAK2 important *geometric operations* like the cutting of geometries in a window (clip) are not executed in the *processing component,* but directly in the graphical kernel system. This means that only few important geometric algorithms are running in the processing component. The reason for that is that the generation of maps as the main functionality of THEMAK2 is only needed for the visualization of the maps. Thus it can be de-coupled from the data handling component. In the *data handling component* elementary data handling tasks like the storage of a file are executed directly with the *file system*. The management of the graphical tools is done by the *graphical kernel system.*

user interface	
geometry operations	processing component
graphical kernel system	data handling component
file system	

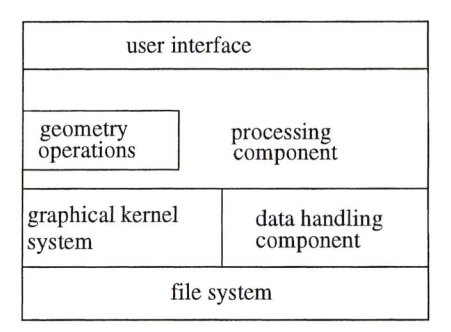

Fig. 5.2 Architecture of THEMAK2 (from: WATERFELD and BREUNIG 1990)

Replacement of the data handling by the DASDBS Geokernel

To replace the original data handling that was realized by files, a physical scheme has to be designed in the DASDBS Geokernel. THEMAK2 distinguishes between G-objects (geometry objects) und A-objects (alphanumerical objects). The G-objects are distinguished so that they can be in relationship with their key to several A-objects. Each A-object has a key (reference to the G-object) and a file as attributes, in which the A-object is stored. Additionally, G-objects can be labelled with a text. The scheme of the NF^2-data model enables the nested representation of G- and A-objects in a single relation. This means that the join, which is necessary in the THEMAK2 system, is already materialized in the DASDBS geokernel.

"G-Objects (Key,
          Segmentgeom,
          Polygonboundary (adr_G-Objects),
          Pointgeom,
          Text,
          A-Objects (Filename,
                    Attributes (Name, Value)))"

A *G-Object* consists of the following attributes: a key, a set of segments, the polygon boundary with reference to the G-Object, the set of the single points, a text and a set of corresponding A--Objects. In the last sub-relation of the scheme above, additionally to the values, the names of the attributes have been stored. Cartographic objects often contain a fluctuating number of attributes. That is why a large number of non reserved attribute values (null values) has to be avoided. Thus the sub-relation consists of name-value pairs. To achieve fast access from a polygon to the segments of its boundary the additional relation can be defined:

"Polygons (Key, Boundary (adr_Segments))"

In the DASDBS geokernel a B*-Tree has provided the fast access to the key attributes. The access to large sets of geometries has been realized by spatial access methods like the Grid File (DOERPINGHAUS 1990) or the R-Tree (BREUNIG 1989). They also can take over the clipping of the geometries at the query window so that this task no longer has to be executed by the graphical kernel system.

In THEMAK2 the management of the files is done by the data handling component with the reading and writing access of one record per file, respectively. Such an access can directly be transformed into one reading or one writing operation of the geokernel. The special balanced tree for the access to key attributes in THEMAK2 can be replaced by the B*-Tree of the geokernel.

In THEMAK2 the join operations have to be realized by costly loops between G- and A-objects. In the geokernel they can be precomputed, as could be seen in the physical database design described above. Furthermore, with the concept of the set-orientation of DB operations in the DASDBS kernel a set of retrieval and insert operations can be executed by a single set-oriented I/O operation. The performance comparison of WATERFELD and BREUNIG (1990) shows that especially for access to G- and A-objects a significant advantage (more than 50% time saved) could be achieved by the THEMAK2 realized with the geokernel in comparison to the original THEMAK2. Furthermore, window queries have been significantly faster with the use of the spatial access method of the geokernel.

5.1.3 The Object Management System

Similar to the DASDBS geokernel the Object Management System OMS (BODE et al. 1992) has also been developed to integrate application specific data types into an open database kernel. In particular the efficient realization of application specific operations played a central role in the development of OMS. Thus the integration of the data types in OMS is possible at a deeper system level directly in the system internal representation of the database kernel. Therefore time critical functions like geometric operations or spatial access methods can efficiently be supported by the internal storage representations of the database kernel.

5.1.3.1 Architecture

The objects of the application are managed by the object manager of OMS. They are mapped to the types of the type manager (Fig. 5.3). The type manager in which a complex structure can be "stamped" to the storage objects corresponds to the cluster and access manager (AM) of the DASDBS geokernel. The AM loads complex objects of the complex record manager (CRM) with a buffer (so called transfer area) into and out of main memory.

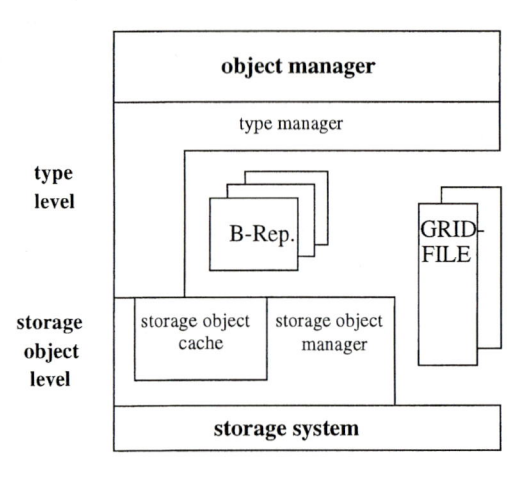

Fig. 5.3 System architecture of OMS[1]

Storage object cache and storage object manager have the task to manage OMS objects efficiently in main and secondary memory, respectively. Examples for the embedding of geometric data types for geo-applications into OMS are geometric objects represented with a boundary representation and with simplicial 2-complexes in three-dimensional space (NOACK 1993; SCHOENENBORN 1993; BREUNIG et al. 1994). Loading the objects from the database, the comparison between the two representations showed that the geometries, which have been implemented with simplicial complexes have a better performance for intersection algorithms. This is true, if the size of the geometric intersection results is small in comparison to the input geometries (BREUNIG et al. 1994). The lower system level of the storage system is comparable with the DASDBS kernel.

5.1.3.2 Deep Embedding of 3D Data Types and Access Paths

With the concept of the "internally embedded types" (IET) OMS can realize new complex types and type constructors directly in the storage object representation. As we have shown in BREUNIG et al. (1994) with the example of 3D objects -which have been represented as boundary representations and simplicial complexes- this can lead to a significant performance gain for geometric operations on top of the OMS kernel. The OMS especially provides the possibility to minimize the transformation costs by choosing the size of the units for the internal database re-

---

1. Modified from: (BODE et al. 1992).

Chapter 5. Systems Development 109

presentations. For example, the specified partial geometries of a simplicial complex can exactly be mapped into the internal representation. Furthermore, spatial access methods like the Grid File (NIEVERGELT et al. 1984) can directly use the storage object representation of OMS.

5.1.3.3 Database Query Language

OMS provides an elementary and strongly typed "basis model language". Its expressions can uniquely be mapped to the types of the type manager. An OMS storage object can consist of hierarchically structured collections of untyped, variably long byte strings. For example, a simplicial complex can be realized in OMS as a storage object list in which important components like the bounding box and the geometry and topology can be stored separately to realize an efficient spatial access (BREUNIG et al. 1994). We give an example that shows clearly the power of the basis model language:

"SELTRANS [RANGE o : inside (o, Box1) AND
    outside (o, Box2), INSERT (ResultBag, o)] (GeoObjects);"

The SELTRANS-operator used in the query above is a general form of the selection and projection operators (BODE et al. 1992) of the Relational Algebra. It applies a transformation to all elements that meet the specified predicate. If the predicate is met, a copy of the result is transferred into a so called result bag.

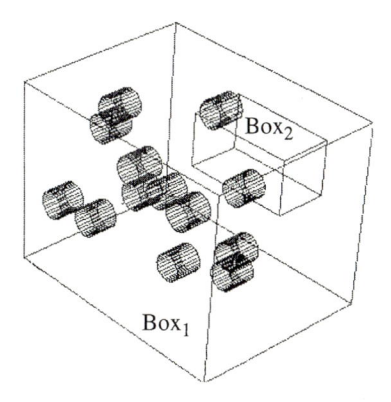

Fig. 5.4 3D-query result in OMS (from: BREUNIG 1996)

In the example query above all geo-objects are selected that are completely inside *Box1*, but outside *Box2*. Figure 5.4 shows the graphical output of the query result.

5.2 The GeoToolKit

In contrast to the DASDBS Geokernel and to OMS the GeoToolKit (BALOVNEV et al. 1997a) which has been developed within the SFB 350 project (NEUGEBAUER 1993) at the University of Bonn, has been designed to support 3D applications from the outset. The toolkit approach exceeds the idea to embed geometric data types and access methods. It also concerns components for the communication of external software systems and for the visualization as part of its user-interface. The component-based approach simplifies the extension of further specific geoscientific components to a 3D/4D geoinformation system.

5.2.1 Historical Development

GeoToolKit is based upon the experiences that have been gained during the development of GeoStore (BODE et al. 1994), an information system for the management of geological data. The pilot version of GeoStore has been implemented on the basis of the relational DBMS ORACLE. However, soon serious problems occurred being caused by the "impedance mismatch" between database representations and, for example, the representation needed for efficient geometric computations. That is why GeoStore was re-designed on object-oriented database technology, first on top of ONTOS and since 1995 on the basis of the OODBMS ObjectStore. Since 1994 the system has been intensively used by geologists[1] as a database support for the interactive geological modelling.

After three years of development work it became clear that GeoStore was too closely oriented to the special geological application described in SIEHL (1993). It could not be used for arbitrary 3D applications. The subsequent analyses of related geoscientific domains showed that to a high degree they share the same functionality that primarily had to do with spatial 3D objects. This was the starting point for the development of a class library as an alternative to the "from scratch" implementation of GeoStore-like applications for concrete domains.

5.2.2 Component-Based Architecture

GeoToolKit is not a GIS-in-a-box package - it is rather a library of C++ classes that allows the incorporation of spatio-temporal functionality within geo-applications. Thus it is primarily oriented on software engineers with C++ experience involved in the development of special purpose geo-applications which can hardly be modelled within a standard GIS. Being a component toolkit, it encourages the development and deployment of reusable and open software.

Figure 5.5 shows the architecture of GeoToolKit. It is subdivided into two main components: a C++ class library and interactive tools. The class library consists of classes for the spatial data handling based on ObjectStore. We will enter into these classes in the next chapter. The graphical classes for the visualization provide the presentation of 2D maps and 2D and 3D regions, respectively. They are based on the object-oriented graphics library CGI-3D (FELLNER et al. 1993). To realize the communication of the GeoToolKit with external geoscientific software systems, first a protocol has been implemented which is based upon UNIX sockets. An advanced system architecture based on CORBA will be introduced in chapter 7.4.

The geodatabase browser belongs to the interactive tools of GeoToolKit. With this browser the geoscientist, for example, can load databases of different examination areas. The navigation through different data sources is also possible and special information about the objects is provided. Furthermore, the geoclass editor will enable the geoscientist in future to generate her or his own geoclasses. Existing classes can be tailored for new requirements. The geoscientist then is in the situation to extend geospecific classes without C++ or detailed database knowledge. The new classes can be placed in relation to other already existing classes.

---

[1]. Working group of Agemar Siehl, Geological Institute, Bonn University, Germany.

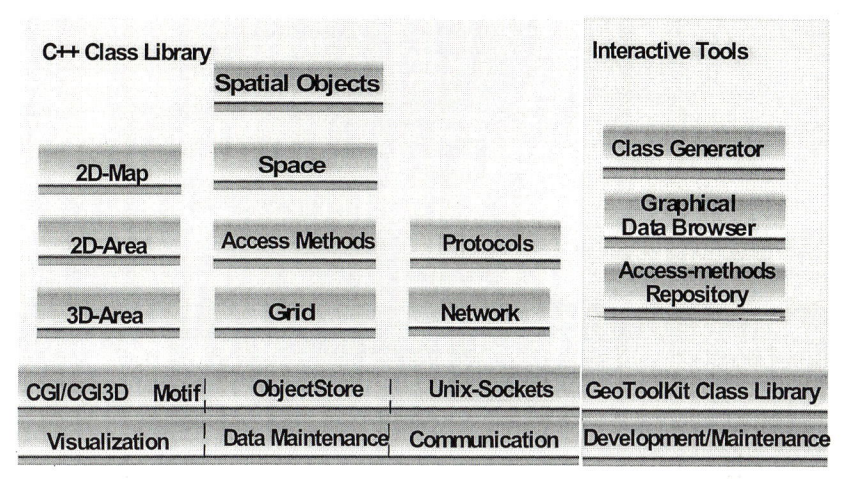

Fig. 5.5 Architecture of GeoToolKit

The inheritance of the geometric functionality in the classes of GeoToolKit enables the geospecific extension of these classes with new attributes and operations. For example, a geological stratum, which is inherited by the GeoToolKit class *Solid* can be extended by the new derived attribute "density" and the new operation *"compVolume"* which computes the volume of the strata. During the insertion and the update of existing attributes and operations the consistence of newly generated or extended classes with all related classes has to be guaranteed. For example, during the update a fault, also the volume of all faults that are stored in the database and the volume of all strata that are related with the updated fault have to be checked automatically. Finally, the access path repository, which is still in development, provides the selection of different spatial access methods.

5.2.3 Data Model

Geo-applications distinguish themselves considerably in their dimensionality, in the type of space (2D/3D), in time (discrete/continuous), in their representations (e.g. raster/vector) and in the way spatial objects and operations are supported. The spectrum of the spatial operations can be extremely large and their implementation often is very costly. Thus GeoToolKit restricts itself to such spatial objects that are used and needed for the interactive geological modelling (SIEHL 1993; ALMS et al. 1998). The interactive geological modelling is inherently three-dimensional. The majority of geological entities (e.g. strata, faults etc.) are preferably modelled as triangulated surfaces in 3D space or as tetrahedralized bodies. These objects are extended by "thematic", i.e. geologically specified attributes and relationships (MALLET 1992a; SIEHL 1993; BREUNIG 1996). The most important spatial operations are the intersection of surfaces and bodies and the sections implemented by horizontal and vertical intersections with a plane.

GeoToolKit's geometric 3D classes build a complete set of simplexes, complexes and analytical 1D-3D objects in three-dimensional Euclidean space. The complexes are homogeneous collections of simplexes with the same dimension, respectively. Spatial objects of different types can be combined into a heterogeneous collection. This so called *group* can then be treated as a single object.

The class hierarchy of GeoToolKit contains classes for the representation of:

1. 0D-3D spatial simplexes *(point, line, triangle, tetrahedron)*;
2. 1D-3D complexes *(curve, surface, solid)*;
3. Container objects *(group)*;
4. Analytical objects *(line, plane)*.

Application specific data types (real-world data types) can either be deformed with a built-in--type of GeoToolKit or with specialization or aggregation of a built-in-type and the composition of such types, respectively.

A completely new data type can be generated, if it is defined as the specialization of the most general built-in-type *SpatialObject*. *SpatialObject* defines an interface for spatial predicates, functions and operators as well as for several service functions that are needed for all built-in--types and user-defined types. Each concrete spatial object has to redefine the methods of this interface depending on the structure and the special requirements.

As every spatial object is surrounded by a bounding box, *SpatialObject* for example declares a function that returns a bounding box. This function then has to be redefined in each direct derivation of *SpatialObject* to return the bounding box of the corresponding object. The special class *Triangle*, for example, has additional functions to check the spatial integrity of the triangle, for the efficient spatial access and for the computation of the normal vector. Finally, it is determined whether objects are in parallel or orthogonal to a *line* or to a *plane*.

The *SpatialObject* interface can be subdivided into the following four categories of operations:

1. *Spatial predicates:* functions that check, if a special unary or binary spatial relationship is true or false. They return a boolean value.
2. *Spatial functions:* unary or binary functions that compute new data on objects or on pairs of objects. They return objects of a type that are not a member of the *SpatialObject* hierarchy.
3. *Spatial operators:* see spatial functions, but they return new spatial objects, i.e. members of the *SpatialObject* hierarchy.
4. *Non-geometric internal service functions:* functions that do not belong to the three categories above. They return non-geometric information and execute non-geometric actions.

We give an example for the implemented spatial predicates, spatial functions and spatial operators of GeoToolKit, respectively:

1. *Spatial predicates:* intersects tests, if two spatial objects are intersecting each other (touching also is treated as an intersection).
2. *Spatial functions:* distance computes the minimal distance between two spatial objects.
3. *Spatial operators:* intersection determines the intersecting geometry of two spatial objects. The result is a new spatial object.

The approach of the hierarchical geometric class library provides the following preconditions:

1. It is guaranteed that each spatial type within the *SpatialObject* hierarchy (built-in and user-defined) provides at least the functionality of the most general spatial object.

Chapter 5. Systems Development

2. As each method that is specified in *SpatialObject* can be applied to each specialization of *SpatialObject*, even *groups* that contain objects derived from *SpatialObject* can be operands of these methods, because *Group* is also a specialization of *SpatialObject*. The grouping of the objects allows to organize data without loss of spatial functionality according to higher abstraction criteria. The derivation tree of *SpatialObject* is functionally closed.
3. The *SpatialObject* hierarchy can be extended by user-defined derivations without loosing the closeness.

With the virtual functions mechanism of C++ it is guaranteed that geometric operators can also be executed without knowledge of the detailed types. The suitable operator is automatically selected.

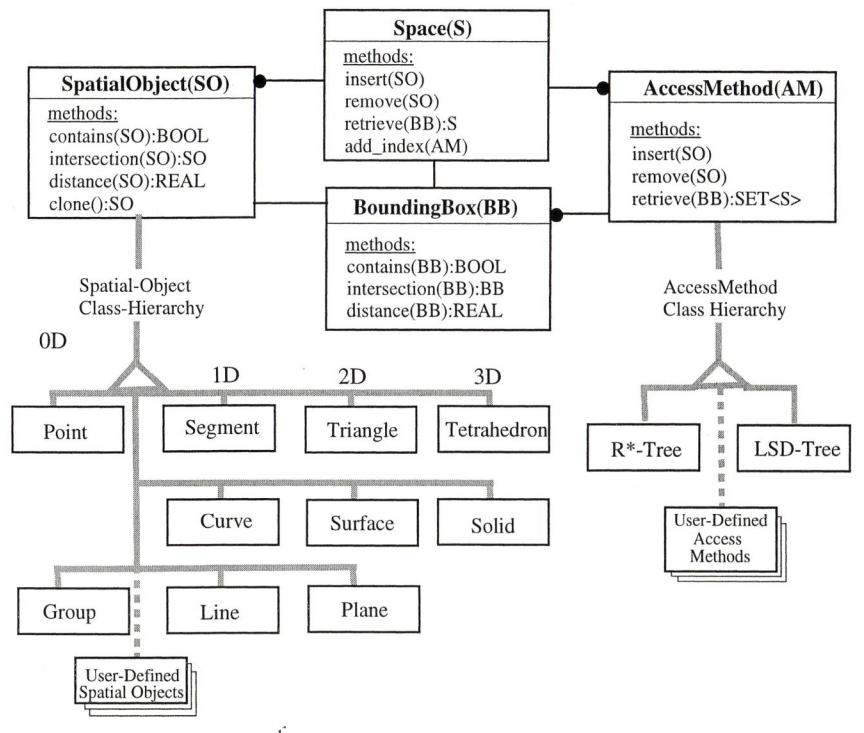

Fig. 5.6 Data model of GeoToolKit[1]

Figure 5.6 shows the relationships between the central classes of the GeoToolKit data model in an OMT-like notation (RUMBAUGH et al. 1991). The object-oriented data model mainly serves as common medium for the communiction between geoscientists and computer scientists. The graphical presentation in a clear object diagram has been proved to be very worthwhile. At the implementation level the OMT data model is mapped to the C++ class library, i.e. to the object-oriented database.

---

[1]·Taken from: (BALOVNEV et al. 1997a)

A *Space* is a special object container class for sets of spatial objects that takes care for the efficient retrieval of its elements according to their position in space, i.e. exactly in space or with a spatial interval. Spatial objects can be inserted into a space (*insert* function) or again be removed (*remove* function) or the user can search for a set of spatial objects (*retrieve* function). Let an interval in the 3D cartesian coordinate system be defined as a cube with axis-parallel surfaces *(BoundingBox)*. The class *Space* serves as the main root point *(add_index)* for spatial access methods *(AccessMethod)* on the embedded objects *(SpatialObject)*. To enable an efficient retrieval, a *space* should at least have one index. The class *Space* can be extended in that way that is also responsible for the transaction management during the access to the spatial objects. With the spatial objects the spatial predicates (e.g. *contains*), spatial functions (e.g. *distance*) and spatial operators (e.g. *intersection*) described above can be executed.

5.2.4 Spatial Representations

At the conceptual level a spatial object (*SpatialObject*) can be defined as a set of points. In practice, however, applications only use a reduced sub-set of spatial objects like line segments, triangles and their combinations. All operations are, as described above, re-implemented for each class under consideration of their concrete representations. A concrete object is modelled as a specialization of a single abstract *SpatialObject* class. *SpatialObject* defines an interface (spatial operations) and a concrete class provides a suitable representation for the object and for the implementation of the functions that are defined in the abstract *SpatialObject* class. A class can only add its inherent functions in these concrete class functions.

GeoToolKit allows the choice between different spatial representations for the adequate use in different 3D applications. Solids, for example, can be represented as simplicial 3-complexes (tetrahedron networks)[1]. A single tetrahedron object of the class *Tetrahedron* is defined with:

1. Its four vertices P_0, P_1, P_2 and P_3;
2. The four normal vectors of the surfaces of the tetrahedron that are determined by the order of the points showing outside.

In this case the simplicial 3-complexes are represented by class *TetraNet* that is derived by the abstract class *SpatialObject*. It contains the following data elements:

1. An R-Tree to store the 3-simplexes of the class *TetraNetElement*;
2. A counter for the number of the 3-simplexes;
3. The bounding box of the tetrahedron network;
4. A flag for the last error that occurred.

The functions for the allocation of persistent storage provided by ObjectStore guarantee that the 3-simplexes of a simplicial 3-complex are stored close together in the database, i.e. in the same segment. A spatial clustering of different tetrahedron networks or of neighboured tetrahedra of a single tetrahedron network, however, cannot be provided.

The 3-simplexes that belong to one simplicial 3-complex (Fig. 5.7) can be represented by the class *TetraNetElement* which contains the following data members:

---

[1]. The corresponding classes have been implemented by HEINKE (1997).

1. Pointer to a *Tetrahedon;*
2. Pointer of the four neighbouring tetrahedra N_0, N_1, N_2 and N_3;
3. A unique identification *(index)* for the tetrahedron (3-simplex).

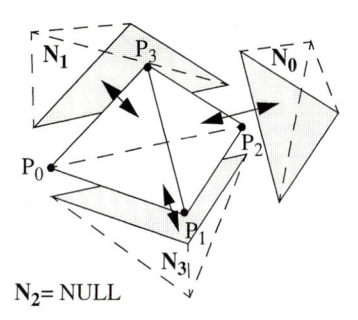

Fig. 5.7 Representation of a 3-simplex within a simplicial 3-complex

Depending on the representation above we define a *valid tetrahedron network* as follows:

1. No 3-simplex is allowed to occur multiply;
2. Two 3-simplexes do not intersect at all or they are touching completely in one of their surfaces, respectively (three common vertices);
3. It is allowed that the tetrahedron network consists of a set of non-connecting components.

The correctness of the tetrahedron network is checked during its instantiation. Alternative representations for geologically defined bodies are the generalized maps and the α-shapes, both introduced in chapter 4.2.1. They are both special kinds of the simplicial 2-complexes and could be embedded into GeoToolKit.

5.2.5 Spatial Indexes

To realise efficient spatial database queries, it is not sufficient to implement geometric data types efficiently in the database. Additionally, the access to the geometric 3D objects for large data sets must be guaranteed.

The research in the field of spatial access methods has developed rapidly (GAEDE and GÜNTHER 1998). Although the R*-Tree (BECKMANN et al. 1990) seems to be one of the most used access methods, hitherto no spatial access method has had a significant advantage over the others for the general case. Against that the one-dimensional case is dominated by the B-Tree (BAYER and MC CREIGHT 1972). For a *Space* in GeoToolKit it is useful to support more than one spatial access method alone. Relevant for GeoToolKit is the extension of a space with new access methods. A space can also possess more than one spatial index. The user can explicitly specify which index should be used for the retrieval. A spatial index can be added and deleted at any time. If no index is associated with the space, the spatial objects are sequentially iterated in space.

The existence of spatial indexes does not exclude the use of non-spatial indexes. The indexes of ObjectStore can also be defined over spaces as if they were usual ObjectStore collections. Non-spatial queries are formulated according to the syntax of the ObjectStore query language. It is also possible to combine spatial and non-spatial indexes within a single space. In this case the spatial and the thematic part of the query are sequentially executed. First the query manager executes the index supporting part of the query. The rest of the query is executed on the result set. If both parts of the query have an index support, the query manager executes them independently of each other and builds the intersection of the result sets. A query optimization according to the order of the executed queries is not yet realized in GeoToolKit. A combined spatial and non-spatial access is supported by multi-dimensional indexes like the LSD-Tree (HENRICH et al. 1989). With such multi key queries the procedure is similar to the pure spatial queries. The spatial query predicate, however, i.e. a three-dimensional box is replaced by a multi key. The user then has to provide a function that converts a spatial object into a multikey specialization class. The index maintenance has to be activated explicitly for updates of non-spatial attributes that are part of the multi key.

In the contrast to the Relational Data Model in which a tuple is exclusively in the possession of the relation, an object can be used shared from several collections (here: *Spaces*). The object can even exist independently of object collections. Updates influence the index structures of all collections in which the object is participating. A spatial object handles a list of all spaces it includes to hold the indexes consistently. Each update that changes the bounding box of the object leads to a maintenance of the index structures in all spaces. As the reconfiguration can be a costly process, the index support should be switched off before serial updates and only be switched on again when the last update has been executed.

The OODBMS ObjectStore provides an advanced, but low-level clustering control mechanism. In combination with the transaction control this makes ObjectStore into a suitable tool for the implementation of different access methods.

The question arises how a newly implemented method is made known to an arbitrary space. Most spatial access methods use the bounding box approximation of spatial objects, because in the cartesian coordinate system spatial operations for bounding boxes are more efficient than on the original geometries. With the assumption that each spatial object has stored a bounding box or that it can easily be computed, the interface of a spatial access method can be reduced to three basis operations: *insert, remove* and *retrieve* for a spatial object and a set of spatial objects, respectively. An arbitrary user-defined access method can be made known to a space with the inheritance of the interface of the abstract *AccessMethod* class. Following this approach, an R*-Tree is implemented in GeoToolKit as one of the embedded indexes. The R*-Tree (BECKMANN et al. 1990) supports the following types of spatial queries:

1. Return all spatial objects that are completely inside a query box (contains query);
2. Return all spatial objects that intersect a query box (intersect query);
3. Return all spatial objects that contain a specified 3D point (point query).

Figure 5.8 shows the spatial distribution of the first 150 and 500 triangles, respectively for a geological stratigraphic boundary of the Lower Rhine Basin, consisting of 1185 triangles. Per node a minimum of 4 and a maximum of 8 entries have been inserted.

Chapter 5. Systems Development 117

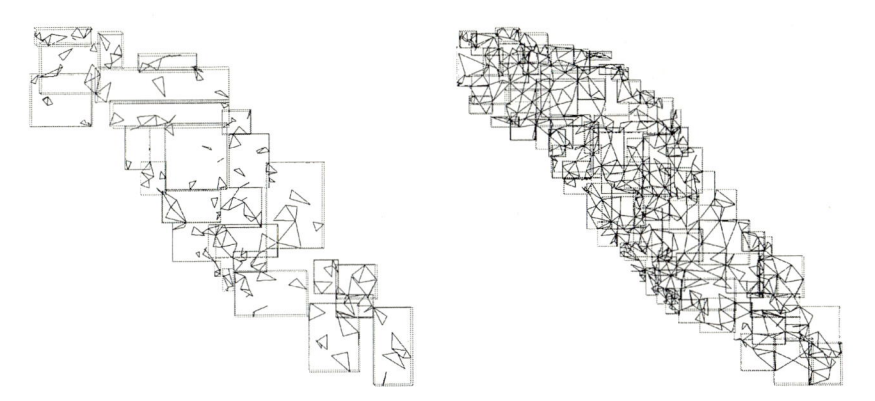

Fig. 5.8 Spatial distribution of the first 150 and 500 triangles, respectively of a stratigraphic boundary surface in the Lower Rhine Basin, managed by a R*-Tree

In the left figure the base structure of the R*-Tree can already be seen clearly. The storage of more triangle networks in the right figure, however, leads to an increasing overlapping of the minimal circumscribing cubes (MCCs)[1]. Tests with geological data have shown that a completely different structure of the MCCs arises, if the insert order is modified during the first insertion steps. The final result after the insertion of all triangles, however, does not show significant differences (WAGNER 1998).

Figures 5.9 and 5.10 show examples for the retrieval in an R*-Tree with a query box and a deletion operation on the same stratigraphic boundary surface of the Lower Rhine Basin, represented as a triangle network.

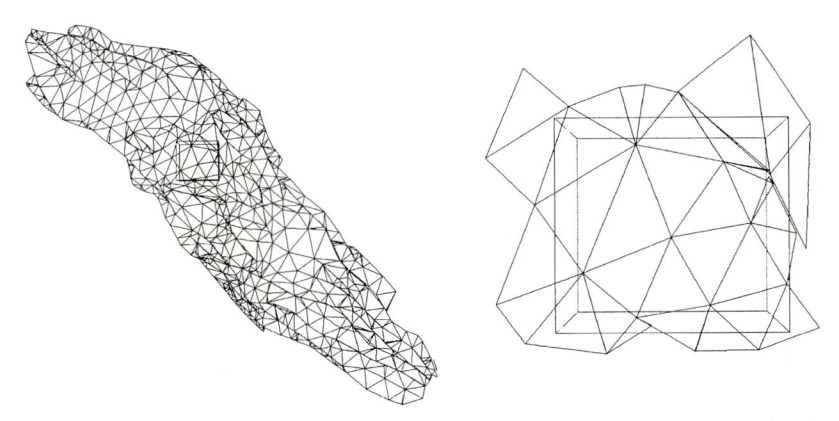

Fig. 5.9 Stratigraphic boundary surface represented as triangle network with query box and retrieval result

---

[1]. In the figure only two dimensions (x-y-plane) of the MCCs are visible. The extension of the geological data in the z-direction, however, is significantly smaller.

The interface of the R*-Tree in GeoToolKit is defined with the *template<class T>class gtRtree*. The constructor of this template class is called with the type information which is stored in the tree. The retrieve-function supports also the "contain" as well as the "intersect" queries, i.e. as query result either the bounding boxes are selected that are completely inside the query box or those that additionally intersect the query box.

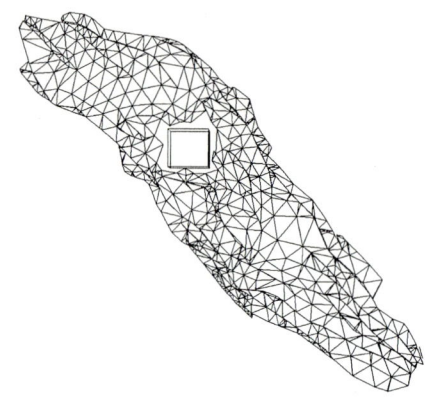

Fig. 5.10 The triangle network from Fig. 5.9 after a delete operation

The R*-Tree in the GeoToolKit is also internally used to accelerate some geometric 3D operations like the intersection of simplicial complexes (BREUNIG 1998). In the first step of the algorithm those simplexes are intersected with each other that are inside the "intersection bounding box". This box arises from the intersection of both bounding boxes of the two intersecting simplicial complexes. Each simplicial complex (triangle network and tetrahedron network, respectively) is stored in their own R*-Tree.

In chapter 4.3.4 we have already introduced the principle of perspective region queries. They support the 3D visualization and animations. The R*-Tree implemented in GeoToolKit realizes the perspective retrieval with three distance zones. For each zone one view cone is provided. So with the visualization component of GeoToolKit large data sets also can be visualized efficiently from different views.

Chapter 5. Systems Development 119

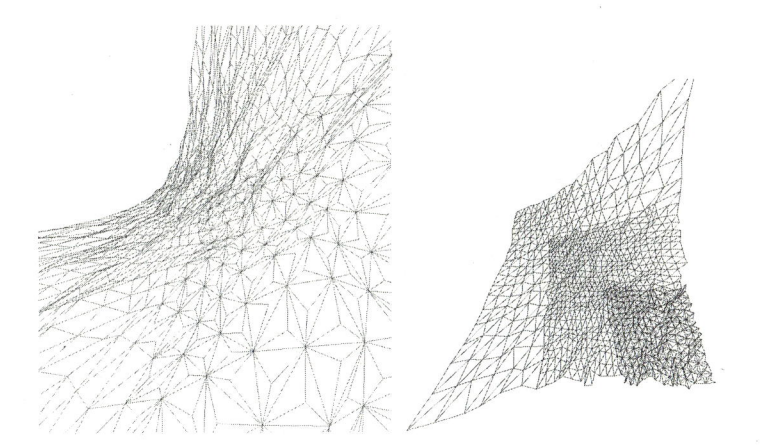

Fig. 5.11 Result of a perspective region query with three distance zones, seen from the view of the observer shown with a stratigraphic boundary surface of the Lower Rhine Basin

In comparison to "usual region queries" the perspective region query in the R*-Tree additionally considers a reference point with view direction and a view opening angle. Figure 5.11 shows the result of a region query composed into three distance zones. Geologically the triangle network of Fig. 5.11 belongs to a stratigraphic boundary surface of the Lower Rhine Basin. The objects, which are in the respective distance zone, have to be loaded from the database as soon as the position of the observer is changing. A copy of the visualized data is given to a rendering machine which computes the transformation into the two-dimensional plane of the picture. Hidden surfaces are determined, the light sources and shadows computed and the colour values transferred.

5.2.6 Extensible 3D Visualization

In principle two approaches can be distinguished concerning the visualization in geoinformation systems. The first approach uses external VRML capable viewers[1] like CosmoPlayer. They use a kind of graphical language like VRML as input. The GeoToolKit classes then only have to write the contents of a file with the corresponding format. After the loading of the objects the viewer works independently of the database. Figure 5.12 shows a 3D visualization of stratigraphic boundary surfaces and fault surfaces of the Lower Rhine Basin generated with CosmoPlayer and exported from GeoToolKit.

---

[1] http://www.sdsc.edu/vml/

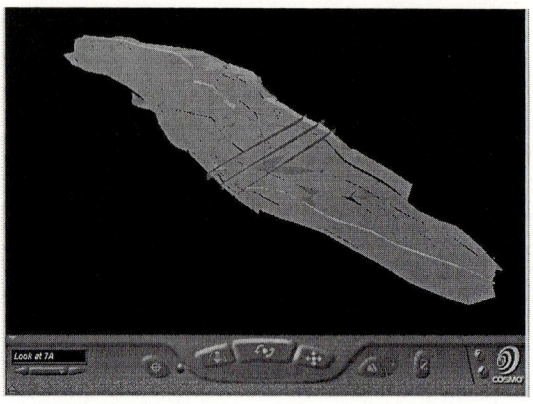

Fig. 5.12 VRML-visualization of stratigrapic boundary surfaces and fault surfaces from the Lower Rhine Basin exported from GeoToolKit

Although a 3D viewer usually provides different ways to manipulate objects (move, rotate, zoom, select etc.), in this solution an interactive communication with the database is nearly possible. Furthermore, most viewers are closed regarding extensions and they do not allow to integrate application specific control mechanisms.

The second approach leads us to the use of programme interfaces like graphical libraries. With this solution the library is loaded to the database client. The developer is completely responsible for the implementation of all control mechanisms. They allow to implement full interactive communication with the database. The best known representative of such a programme interface is OpenGL. The disadvantage of OpenGL-similar solutions, however, is that they only provide a low-level interface. More interesting are object-oriented high-level interfaces like GRAPE (1997) or CGI-3D (FELLNER et al. 1993). They allow the operation of objects as they are defined in GeoToolKit.

In GeoToolKit both approaches of the visualization are realized. There is also a VRML interface as its own 3D visualization component which is implemented with CGI-3D. The CGI-3D is a C++ class library for the representation of three-dimensional objects. It is implemented on the basis of OpenGL and profits from the hardware support. CGI-3D is realized as a C++ library; the integration with GeoToolKit is straight forward. Each GeoToolKit class is attached with a special function that converts the respective object into the CGI compatible class. Figure 5.13 shows the 3D visualization of selected horizons from the Lower Rhine Basin realized with CGI--3D.

The 3D visualization component of GeoToolKit also provides some high-level classes *(3DArea, 2DArea, 2DMap)* that combine the management of spatial objects in a window with elementary control mechanisms. For the implementation of the window management and for the control mechanisms OSF Motif has been used. A Motif-based GUI-builder has been incorporated to develop graphics interfaces efficiently. It generates the C++ Code of the graphics interface. Thus the toolkit idea is consequently carried on to the user-interface.

Chapter 5. Systems Development 121

Fig. 5.13 Visualization of selected horizons and fault surfaces from the Lower Rhine Basin realized with CGI-3D

In GeoToolKit there is also the possibility to visualize moving objects. Therefore a coupling with the 3D visualization tool GRAPE (1997) on the basis of UNIX sockets is realized. We will enter into this point in the next chapter.

5.3 Example of an Application: Balanced Restoration of Structural Basin Evolution

Some ideas of the balanced restoration of structural basin evolution for the Lower Rhine Basin have been implemented in the first prototype software system "GeoDeform" (ALMS et al. 1998). This application is well suited to demonstrate the spatial and temporal modelling and management of geo-objects. The geological process of the reconstruction can clearly be demonstrated with a computer supported animation and an adequate 3D visualization. However, this animation cannot be handled efficiently without a database supported management for large sets of time-dependent 3D/4D data.

Let us introduce the application (SIEHL 1993; KLESPER 1994) in more detail. It has already contributed substantially to the development of GeoStore (BODE et al. 1994; BREUNIG 1996; ALMS et al. 1998), an information system for the management of geologically defined geometries. Also GeoToolKit (BALOVNEV et al. 1997a) has been designed according to requirements of this application. Figure 5.14 shows a map of the examination area in the Lower Rhine Basin with open cast mines, sections, fault lines and the adjoining mountains.

The basis for the modelling of the Erft block is interpreted lithostratigraphic sections[1] in the horizontal scale 1:10000 and the vertical scale 1:2000. The parallel sections run in the direction SW'NE direction and cover a distance of 1.5 to 4.5 km. In the section supplementary to the fault points and stratigraphic points for each point further information like the stratigraphy, the name of the respective fault or stratum and the topological relationship of the point with neighboured strata and faults are given. These "hard wired" topological relationships of the left and the right fault as well as the upper and the lower stratum line of a stratum point, however, can only be modelled if we start out from the relatively simple geological structures in the Lower Rhine Basin.

---

[1]. Kindly provided by Rheinbraun AG Cologne, Germany.

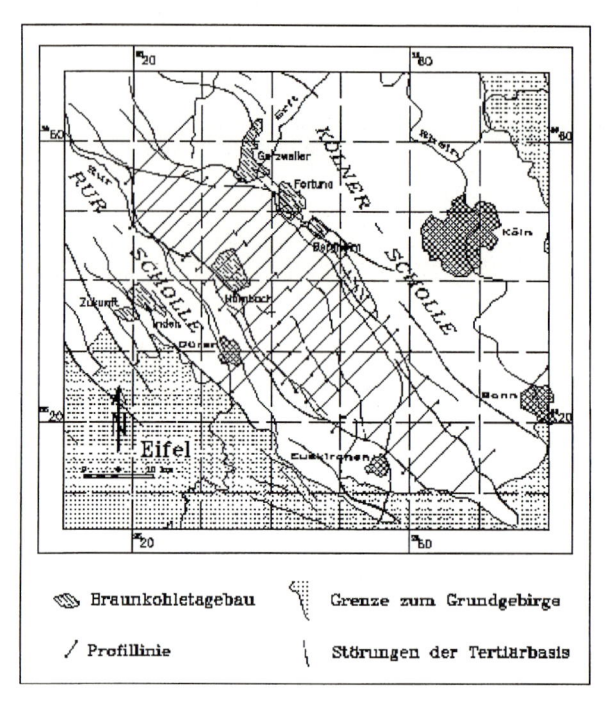

Fig. 5.14 Map of the examination area in the Lower Rhine Basin with the open cast mines, sections, fault lines and the adjoining mountains

The goal of this geological application is to reconstruct a consistent geological 3D model of the Lower Rhine Basin (SIEHL 1993). Finally a model-based balanced geological restoration has to be developed. In constrast to the usual procedure applied else where, as a first step the already interpreted sections (see Fig. 5.15) and not the wells have been taken. The wells, however, have also been available as original data.

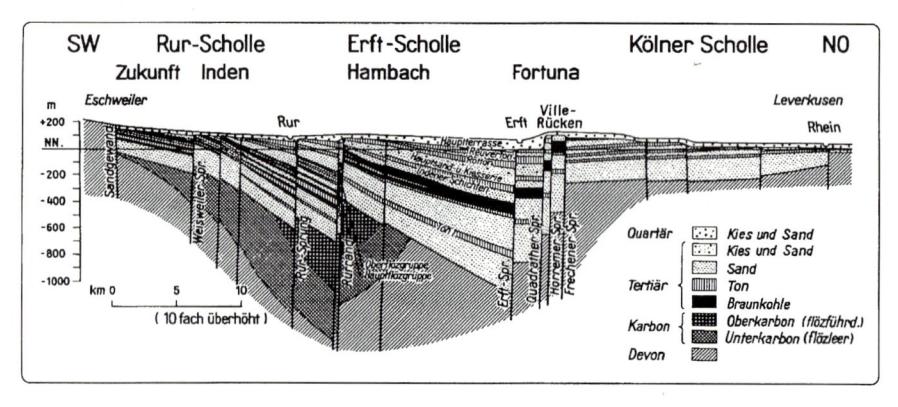

Fig. 5.15 Section in the Lower Rhine Basin (from: RHEINBRAUN 1991)

Chapter 5. Systems Development

By these means the costly step of well correlation could be passed over in the first instance. As we shall see in chapter 5.4.3, however, the well data can, a posteriori, be used to check the sections and the modelled stratigraphic boundary surfaces, fault surfaces and stratum bodies.

In the present application stratigraphic boundary surfaces and fault surfaces have been generated and -if necessary- interactively corrected (KLESPER 1994). The surrounding polygons and the existing isoline plans of the faults have been taken as the basis for the triangulation.

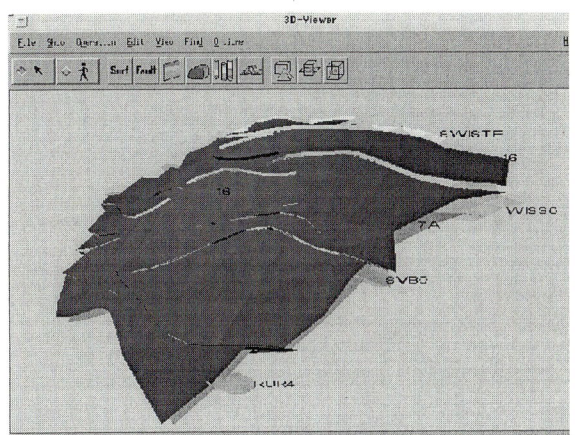

Fig. 5.16 Stratigraphic boundary surfaces and fault surfaces of the Erft block in the Lower Rhine Basin generated between sections

The modelling process which has been applied to the Erft block can be subdivided into the following three steps:

1. Digitalization of geological sections, i.e. the generation of line data from point data;
2. Construction of geological boundary surfaces (stratigraphic boundaries, fault surfaces) on the basis of point sets on the sections obtained during the digitalization, i.e. the generation of surface data by triangulation of line data;
3. Construction of volumes (stratigraphic bodies) from stratigraphic boundary surfaces and fault surfaces.

To minimize the roughness of the surface the triangulation has been refined with the Discrete--Smooth-Interpolation Algorithms DSI (MALLET 1992b). In the following step of the 3D modelling stratigraphic 3D bodies are generated from adjacent stratigraphic boundary surfaces and fault surfaces as well as from the sections. It is useful to execute this step first in a larger scale, e.g. for a single open cast mine to increase the geological meaningfulness of the 3D model.

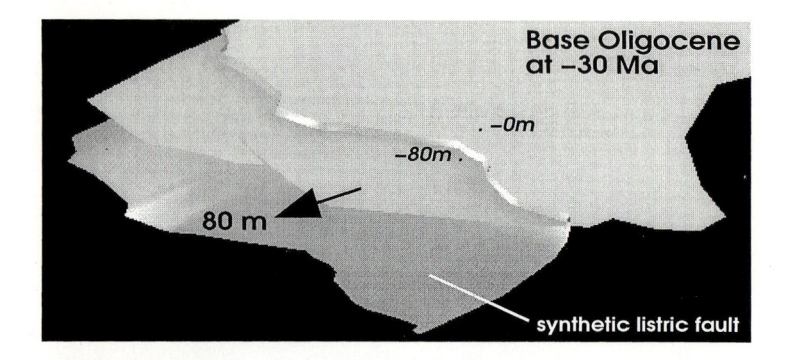

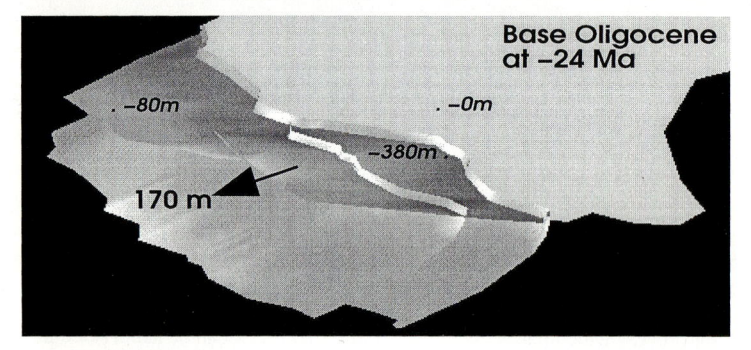

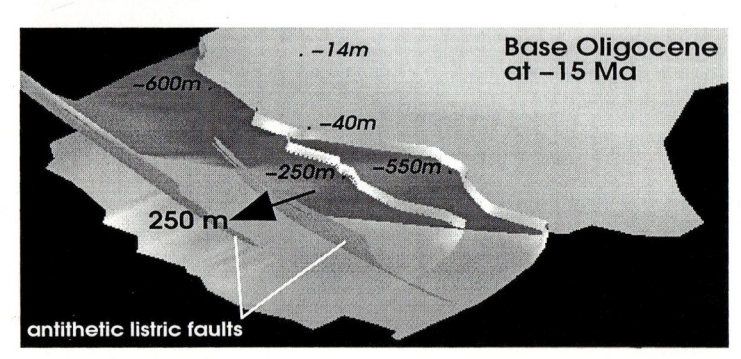

Fig. 5.17 Three time steps of the basin modelling in the Lower Rhine Basin: view from the southern part to northeast with the Oligocene bases and the antithetic faults (from: ALMS et al. 1998)

Figure 5.17 shows three time steps of the basin modelling for the Lower Rhine Basin as it has been modelled in the framework of the balanced geological restoration with the programme system GeoDeform (ALMS et al. 1998). Inter alia the model starts out hypothetically from a rotation of the Eifel mountains relative to the east Rhenish Massif. The centre of the rotation could then be south of Bonn approximately in the Neuwied Basin.

The application requires three types of database queries:

1. *Spatial queries for strata and faults inside a section:* for example, the user can select all strata with a certain upper stratum in a given depth interval and between two specified faults.

2. *Spatial queries for triangulated stratigraphic boundary surfaces and fault surfaces:* this type of queries contains geometric 3D operations like horizontal and vertical sections through the stratigraphic boundary surfaces and the fault surfaces, the intersection of two surfaces and the clipping of a surface with a bounding box.

3. *Spatial and temporal queries with stratigraphic boundary surfaces and fault surfaces:* the spatial queries can be combined with queries for time intervals (like "return all geometries between 6 and 13 million years").

We will enter into temporal queries in chapter 5.4.5.

5.4 Realization of Geological Components for a 3D/4D Geoinformation System

The geoscientific expert is not only interested in a "geometric ToolKit", but requires a toolkit that supports her/his own discipline. In the case of geology the way has to be made from the geometric toolkit to the geological toolkit. First the extensibility of new applications concerns the data model, i.e. its classes and relationships. Furthermore, it is important that integrity checks can be integrated into the data model and into the query language of a geodatabase. Last but not least, new geometric data types should easily be generated by a browser without internal database or programming knowledge.

5.4.1 Object Model Editor

We show the design of a geological object model for the examination area in the Lower Rhine Basin. The object model has been developed in a detailed object-oriented analysis of the different steps of the geological modelling process. The database design typically should be developed by an object model editor like that described in PARENT et al. (1998). In the meanwhile commercially available editors can also be used that translate the classes and their relationships into object-oriented code.

Following the object-oriented modelling technique we require from a geological toolkit that it not only supports abstract geometric primitives like points, curves, surfaces and volume, but that also entities of the "real world" like geological strata, sections and faults are supported. The applications developed with the geological toolkit then inherit as many geometric functions of the GeoToolKit as they need and supplement them with application specific semantics.

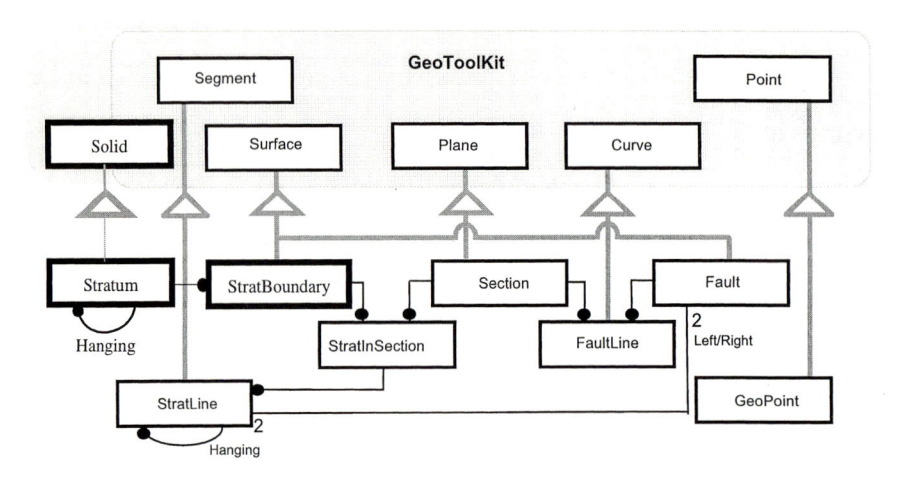

Fig. 5.18 Construction of a geological data model for 3D geometries on top of GeoToolKit

Figure 5.18 shows a geological object model for 3D geometries realized for the Lower Rhine Basin on top of GeoToolKit. The three essential classes in the geological object model are *Stratum*, *Section* and *Fault*. A geological entity like a stratum is represented as a specialization of the corresponding geometric entity *(Solid)*, extended by specific geological attributes (stratigraphy, lithology etc.) and topological relationships like the upper stratum or the left and the right fault of a stratum line within a section. The geological entity inherits the complete geometric functionality of its GeoToolKit parent class.

The volumes are internally modelled as simplicial 3-complexes (tetrahedron networks). This representation allows also a good approximation of irregular geological bodies. It also has the advantage that geometric 3D operations like the intersection of a stratigraphic body with a fault surface can be reduced to relatively simple operations like "tetrahedron intersects triangle".

Other advanced geological examination areas like the North German Basin in Lower Saxony require the management of more complex geometries. Therefore the geological object model has to be extended for example for the modelling of faults (BREUNIG et al. 1999). The geometry of a fault surface in the introduced object model essentially consists of a single connected fault. In the Lower Saxony application mentioned above often one fault consists of a several geometric components.

5.4.2 Integrity Checking Component

Let us consider integrity constraints tailored to the interactive geological 3D modelling (SIEHL 1993). These finally will build the framework for the integrity checking component of a geological ToolKit.

We distinguish the following types of geologically defined integrity constraints to check the consistency of the geological 3D model:

1. Integrity constraints that concern the non-spatial attributes (e.g. lithology);
2. Topological and topological sorting integrity constraints;
3. Geometric integrity constraints;
4. Integrity constraints that check the number of geologically defined objects of one or more classes.

The following integrity constraints are restricted to specific model conditions, they might not be valid for general models. In Fig. 5.19 "intersection points" on sections are explained that play a central role in the integrity constraints below.

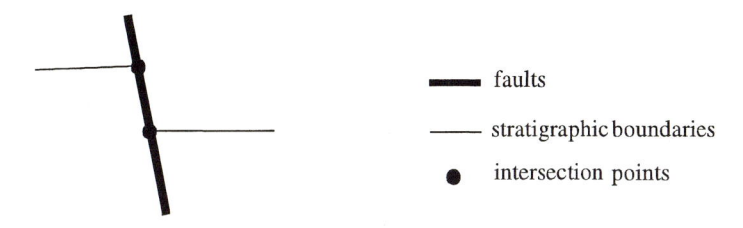

Fig. 5.19 "Intersection points" between stratigraphic boundaries and faults at a selected section

Under these conditions typical examples for integrity constraints of type (1) are:

i. "The stratigrapic sublines of each stratigraphic line must have the same age". Note that each stratigraphic subline corresponds to one section. One stratigraphic line is usually a member of several sections.
ii. "All intersection points of a fault line must have a different lithology".
iii. "The lithology order of the digitalized and of the computed section must be congruent".

Typical examples for integrity constraints of type (2) are:

i. "It is not allowed that any point of the stratigraphic line of a stratigraphic boundary has the same upper stratum as an arbitrary point of another stratigraphic line of this stratum has as lower stratum".
ii. "The end point of a fault line must not have a lower stratum than the previous points".
iii. "All intersection points of a fault line must have the same left and the same right fault".
iv. "For each intersection point P_i of a fault line except the first and the second point and the one before the last, and the last point there is exactly one point P_j, which is different from P_i on the same fault line that has the upper stratum of P_i as its lower stratum".
v. "It is not allowed that two intersection points of the same fault line have the same upper and lower stratum".
vi. "The stratigraphic order of a single fault line must be topologically sorted".

If the stratigraphic sublines are topologically sorted according to their left and right fault and the fault lines are sorted according to the upper and lower stratum, then a cycle-free and connected sequence of all stratigraphic sublines exists. Furthermore, there is a cycle-free and connected sequence of all stratigraphic sublines between two faults and between the upper and lower strata, respectively (HILGER 1998). Figure 5.20 shows part of a protocol file to check the consistency of stratigraphic boundary lines and fault lines of a digitalized section.

```
GeoStore, Version 3.0b2

    GIS to manage 3D geological data
    Computer Science Department III
    SFB 350, University of Bonn, 1993-1999

Integrity checking section: 'vertical intersection 16'
    stratigraphic line: '8_2'
        stratigraphic subline: Lst: ---/RSt: SVBO /lower-
        --> error: left fault line not existing!
        --- upper stratigraphic subline not existing!
        --- lower stratigraphic subline not existing!
        --> error: intersection with right fault line is no endpoint!
            (840.447961, 0.000000, -28.473347)(886.837717, 0.000000, -29.425103)

        stratigraphic subline: LSt: SVBO /RSt: SWISTE /lower: 7A_2
        OK

        stratigraphic subline: LSt: SWISTE /RSt: SWISTO /lower: 7A_2
        --> error: intersection with the left fault line is not endpoint!
            (8708.266672, 0.000000, -322.074979)(9796.434795, 0.000000, -315.326389)
    stratigraphic line: '16_1'
        stratigraphic subline: Lst:SVBO /RSt: WISSO /lower: 8_2
        --- upper stratigraphic subline not existing!
        --> error: no touching with the left fault line!
            (2545957.049540, 0.000000, 40.513994)(5538.324741, 0.000000, 39.193742)

    error: stratum order in the fault lines: 'SVBO' and 'WISSO' is not consistent!
```

Fig. 5.20 Part of a protocol file to check a section of the Lower Rhine Basin

Typical examples for integrity constraints of type (3) are:

i. "Each intersection point of a fault line must also occur as boundary point of a stratigraphic subline".

ii. "The geometry of the digitalized and the computed section must be equal. If not, compute the divergence".

There is an efficient way to check a digitalized section. Viewing the map of the examination area, we can compute a section -as the intersection of a vertical plane with all modelled stratigraphic and fault surface data- exactly at the same place at which the digitalized section is located. We only have to follow the already existing section with the mouse on the screen. This "computed section" is then the reference object to the modelled section. Figure 5.21 shows the comparision of a digitalized and a modelled section with the computed section. By selection on the map the computed section has been positioned exactly at the same location as the modelled section. In the right figure there are less lines, because less stratigraphic and fault surfaces have been digitalized as corresponding lines on the sections. Small divergences of the lines can be due to digitalization errors in the sections or to errors during the modelling of the surfaces between the sections. It is the task of the geologist to verify this in more detail.

Chapter 5. Systems Development 129

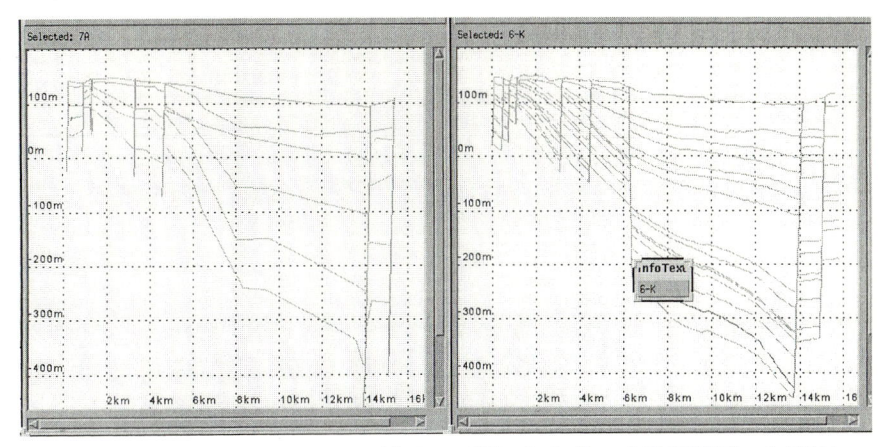

Fig. 5.21 Digitalized and computed section of the Lower Rhine Basin (from: BALOVNEV et al. 1997a)

Typical examples for integrity constraints of type (4) are:

i "The number of the stratigraphic boundary and fault lines on the digitalized and on the computed section must be equal".

ii. "The number of the stratigraphic surfaces must be equal to the number of stratigraphic lines".

iii. "The number of fault surfaces must be larger than x".

Figure 5.22 shows how the stratigraphic subline of a digitalized and of a computed section can be compared with each other. As the measure for the divergence the "enclosed area" of both lines is used.

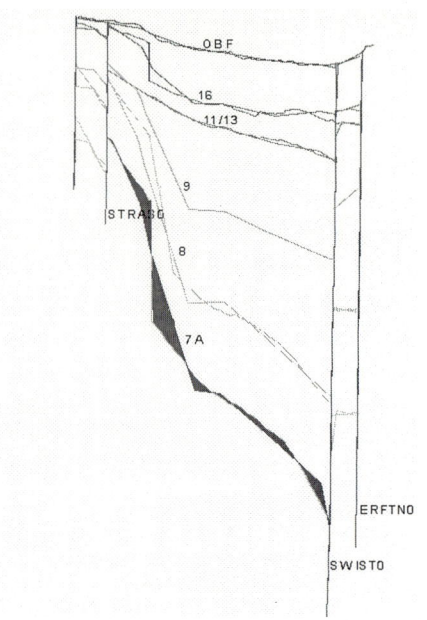

Fig. 5.22 Comparison of two stratigraphic sublines with the measure of the "enclosed area" (black area)[1]

The size of the black area correlates to the measurement of the geometric divergence between the two stratigraphic sublines. Usually a large area also means a large divergence for the geologist. However, the geologist has to decide if a modelling or a digitalization error occurred. For example, a digitalization error of one single point can lead to a large value for the "enclosed area" above.

It is useful to project all wells that are near to a section to keep the lithological order between the wells and the sections consistent. Thus simple geometric divergences of the stratum order between wells and section can be found. The one-dimensional well data are compared with the two-dimensional sections. Figure 5.23 shows the result of such a spatial integrity control.

---

[1]. From: (HILGER 1998).

Chapter 5. Systems Development 131

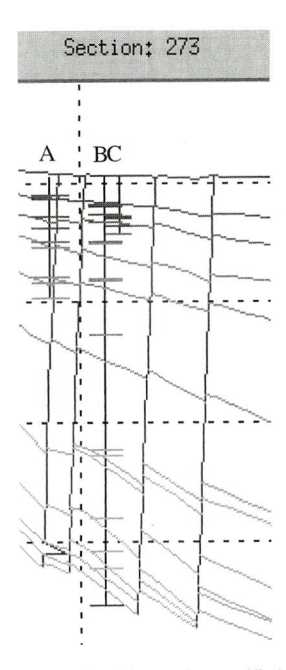

Fig. 5.23 Projection of the wells that are closer than 3 km to the specified section 273.

Figure 5.23 shows the projection of the wells *A, B* and *C* from the database that are closer than 3 km to a specified section. The not exactly vertical lines are faults. The projection of the wells from both sides of the section allows integrity checks between the lithological order of the wells and the sections. More detailed information about the lithology can be retrieved with the database browser. The checking is especially useful to find geometric errors in the original data and classification errors or errors in the model data. The stratigraphy of the wells has to be compared with the stratigraphy of all modelled stratigraphic boundary surfaces, if we also intend to check the lithological order between the sections, i.e. not only on the sections themselves. For this the intersection of a vertical line with all stratigraphic boundary surfaces has to be computed exactly at the start point of the respective well. The corresponding integrity constraint is:

- "The stratigraphic order of the wells should geometrically and topologically agree with the stratigraphic order of the section and the modelled stratigraphic boundary surfaces, respectively".

If the integrity constraint above is not true, we have to check, for example, if a fault is between the well and the section, or if a fault is near the well. It is also possible that an error has been made during the digitalization of the presentation.

5.4.3 Database Support for the Interactive Geological Modelling

A "geological toolkit" especially has to support the interactive 3D modelling of the geological sub-surface. Thus the access to spatial parts of geologically defined geometries has to be provided.

For example, the geologist should have the possibility to "cut" a part of a triangle network or tetrahedron network and to "fit it in" again after the interactive 3D processing. The updates have to be stored in the database. Figure 5.24 shows an example for the cutting and the fitting in after an interactive processing of a cut triangle network (so called *cut-and-paste operation*).

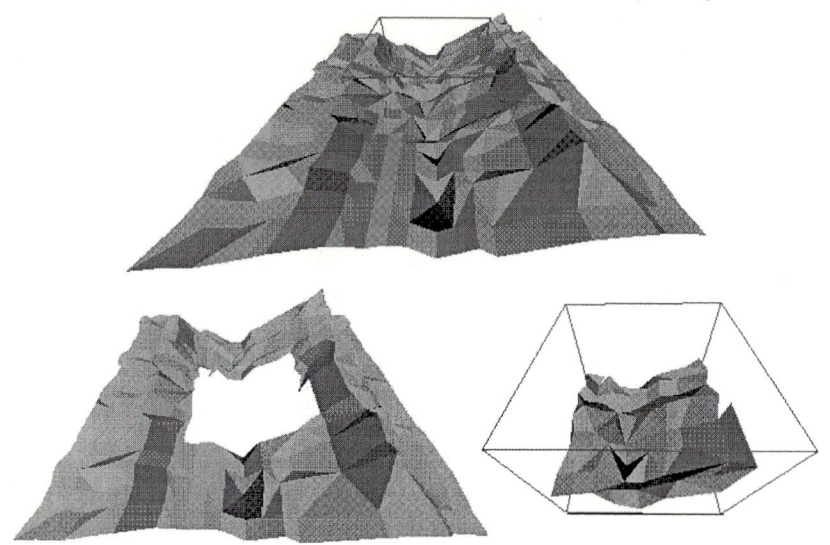

Fig. 5.24 Part of a horizon from the Lower Rhine Basin processed with the cut- and paste operation[1]

Another important operation is the selection of certain geometries of a volume, given from a specified plane *(getPart-Operation)*. Let the volumes be represented as tetrahedron networks. In this operation we specify with a marker, from which side of the plane the result of the operation is taken. Figure 5.25 shows three kinds of tetrahedra that have to be considered in the *getPart*-operation. They distinguish themselves by the way all of them, some or no end points of a tetrahedron are on the same side as has been specified with the marker.

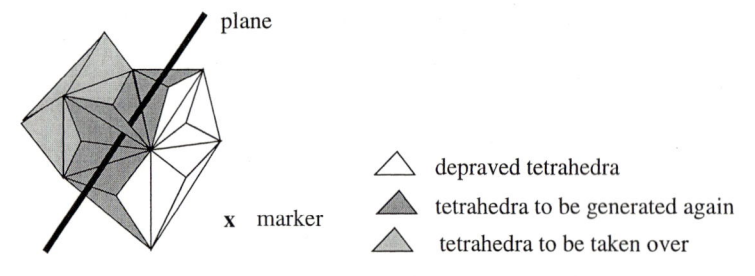

Fig. 5.25 Tetrahedra that have to be considered in the getPart operation[2]

The four different cases that can occur during the intersection of a tetrahedron with a plane are the following to be seen in Fig. 5.26. We assume that the touching at a point, at an edge or at a surface of the tetrahedron is not included when considering the intersection of two tetrahedra.

---

[1] The figure has been produced by Wolfgang Müller, University of Bonn, Germany.

[2] In the figure only triangles instead of tetrahedra have been drawn to achieve a clear presentation.

1. Three intersection points containing no vertex of the tetrahedron;
2. Three intersection points containing one vertex V;
3. Three intersection points containing two vertices V_1 and V_2;
4. Four intersection points.

In the last case (4) no intersection point can be a vertex of the tetrahedron.

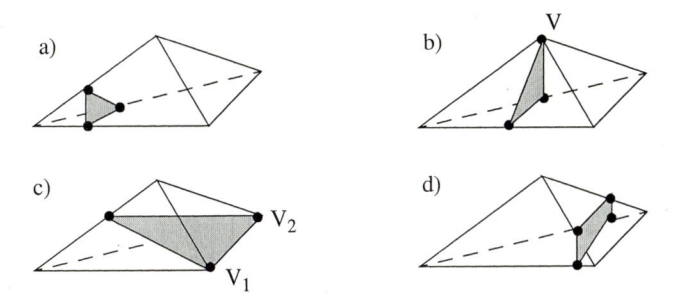

Fig. 5.26 Different cases for the intersection of a tetrahedron with a plane[1]

Figure 5.27 shows the intersection of a plane with sections, fault surfaces and a stratigraphic boundary surface of the Lower Rhine Basin. In this figure the result of the *getPart* operation is shown for both sides of the vertical plane.

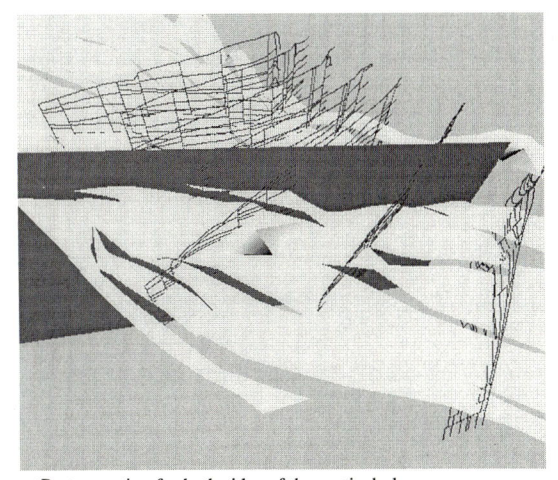

Fig. 5.27 Result of the getPart operation for both sides of the vertical plane

---

[1]. For clarity the grey plane is only drawn inside the tetrahedron.

We give another example of a 3D operation, the *intersection* function of the *TetraNet* class in GeoToolKit which computes the intersection of two tetrahedron networks. We also apply the optimization techniques with bounding boxes, the reduction of the operations to single simplexes and the use of an R-Tree for the access to the single simplexes. Examples for the application of this operation in geology are the intersection of a stratigraphic volume model with a facies volume model or the intersection of a volume model at different times. As the operation is directly executed on the object-oriented database, we have to be aware that during a transaction no updates of the objects are executed. This means that for example the marking of 3-simplexes in a simplicial 3-complex during the execution of an intersection algorithm should not be carried out at the objects themselves. The marking should take place transiently in tables stored in main memory. This means that the object has not to be locked during the execution of the intersection operation.

We reduce the intersection of two simplicial 3-complexes (tetrahedron networks) to the intersection between 3-simplexes; i.e. only the respective intersection between two tetrahedra has to be computed as a basis operation. If we computed the convex hull from the resulting point set in a single step, then the topological information by which edges have been connected with each other would get lost. Furthermore, no constraints could be defined for the tetrahedralization, because the structure of the tetrahedralization is ambiguous. The new edges, however, have to be considered during the tetrahedralization of the corresponding neighbour elements. That is why we execute the intersection of two tetrahedra in the following two separated steps:

1. Construct a graph structure from the intersection of the two tetrahedra (WireFrame);
2. Construct the convex hull from the graph structure.

Algorithm *intersection (T)*

/* Let *T* be a tetrahedron network of type *TetraNet**. Let *bb1* be the bounding box of *this* and *bb2* the bounding box of *T*. Let *this* be stored in an *R*-Tree1* and *T* in an *R*-Tree2*. Compute the intersection between *this* and *T*. Return the set of the intersecting objects as the geometry of type *SpatialObject* */

    carry out the initialisations;

<u>Step 1</u>:

    if [bb1->intersects (*bb2)]   // bounding boxes are intersecting
    {
    // collect the tetrahedra of both networks in a set with a query to the R*-Trees of both networks,
    // respectively

    os_Set<TetraNetElement*>* tetraSet1   = R*-Tree1->retrieve(*bb2,1);

    os_Set<TetraNetElement*>* tetraSet2   = T->R*-Tree2->retrieve(*bb1,1);

    // insert the tetrahedra that are inside the intersection bounding box of tetraSet1 and TetraSet2,
    // respectively into their own R*-Tree R*-TreeErg;

Chapter 5. Systems Development 135

Step 2:

```
TetraNetElement  x = BFS->first();           // first element of breadth first search
while (x)
{
    // run through the elements of this which are inside the intersection bounding box
    to_intersect = T->R*-Tree_erg->retrieve(x->BB, 1);
    // determine for each element those tetrahedra of T_i that intersect the bounding box of the
    // present element
    for all y in to_intersect
        frame = Intersect-Wireframe (x, y);  // intersect the present element with all found
                                             // tetrahedra and construct the graph structure

    for all neighbours n of x
        frame->insert (Constraint [n] );     // tetrahedralize the result
        tmp = convex_hull (frame);           // compute the convex hull
        result->safe_insert(tmp);            // collect the determined intersection objects
        Constraint [x] = frame->new_wires();
        x = BFS->next();
    } // while
} // endif
```

Step 3:
```
if (result == leer)                          // result set is empty
{
    delete result;
    return NULL;
}
        return result;
end intersection (T).
```

In *step 1* it is checked if the bounding boxes of both networks intersect each other. Precondition is that *this* is stored in an *R*-Tree1* and *T* in an *R*-Tree2*, respectively. For networks that consist of several sub-components first the bounding boxes have to be determined. Then the function has to be applied to all bounding box pairs. If the bounding boxes of the two tetrahedron networks are not intersecting each other, the tetrahedron networks also cannot intersect each other and NULL is returned as the result. If the bounding boxes are intersecting each other, the tetrahedra elements of the two tetrahedron networks which are inside the intersection bounding box are collected in a set, respectively. For this, simple queries to the R*-Trees of the both tetrahedron networks have to be executed. After that the elements of the network that are inside the intersection bounding box are stored in an extra R*-Tree. After that it is possible quickly to access certain tetrahedra of the tetrahedron network *T*.

In *step 2* the elements of *this* being inside the intersection box are sequentially run through and for each element those tetrahedra of *T* are determined that intersect the bounding box of the present element. The present element is then intersected with all found tetrahedra and the respectively determined intersection objects are inserted into the result set with the function *safe_insert*.

Finally in *step 3* it is checked if the result set is empty. If this is the case, NULL is returned. Otherwise the determined result of the intersection is returned.

In particular the reduction of geometric 3D operations to elementary geometric operations as we have just shown with the example of the tetrahedron-tetrahedron intersection, has proven to be very useful. This technique leads to a more comprehensible and shorter source code. Run time for the intersection and composition of the result geometries with regard to the number of tetrahedra inside the intersection bounding box in step 2 is still quadratic and can be further optimized.

The well known techniques of computational geometry in $\Re^2$ like plane sweep cannot be applied to $\Re^3$. The complexity of most geometric algorithms in $\Re^3$ cannot be reduced from $O(n^2)$ to $O(n*log\ n)$ as in the two-dimensional case, but can only be reduced to $O(n^2*log\ n)$ (PREPARATA and SHAMOS 1985).

However, we can apply the following optimization techniques for geometric 3D operations like the intersection of two volumes:

1. The reduction of the input data set by using the intersection of approximated bounding boxes;
2. The storage of single geometric primitives in a spatial access method to support the efficient access to sub-geometries;
3. The reduction of the complexity of intersection problems with the use of intersections of elementary geometries like single tetrahedra;
4. Avoiding the generation of unnecessary decompositions, for example if one object is (partially) inside the other and a complete new triangulation is generated for the intersection result.

5.4.4 Support of Different Examination Areas

One of the essential properties of the geological ToolKit is that it is applied in the same way to different examination areas regardless of their special requirements. Figure 5.28 shows windows with a section of the already introduced examination area in the Lower Rhine Basin as well as the map, well data and 3D visualization of the southern part of the North German Basin in Lower Saxony. The application last mentioned will be presented in chapter 6.5.

Chapter 5. Systems Development 137

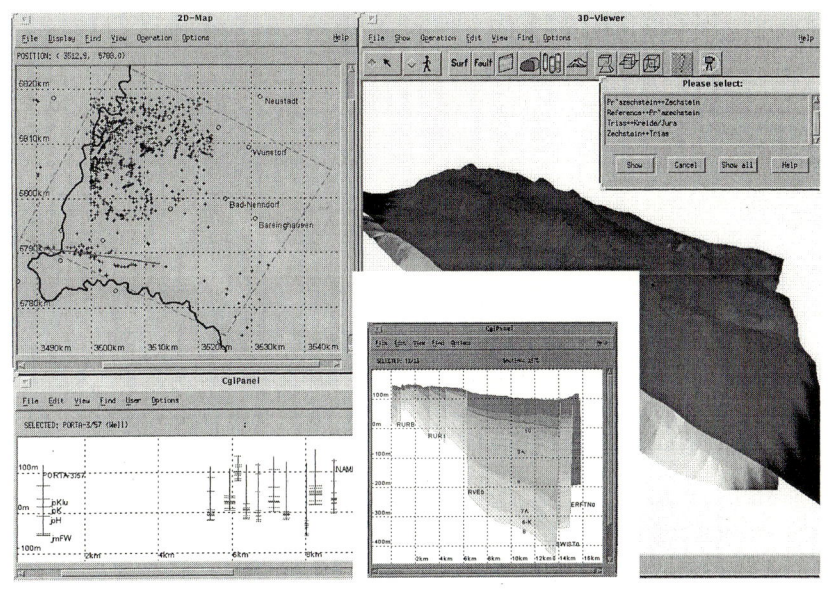

Fig. 5.28 2D and 3D visualization of heterogeneous data and database browser of the two examination areas "Lower Rhine Basin" and "North German Basin - Lower Saxony"

The database browser of GeoToolKit enables to load heterogeneous data sources like wells, digital elevation models, stratigraphic boundaries and faults of the respective application and to visualize them in the map (2D) or in 3D space. Both applications can use the geometric 3D classes, the spatial access methods and the graphics classes of GeoToolKit. Precondition is that the applications do not use new geometric classes. The geological classes on top of GeoToolKit can be extended or changed with the database browser.

5.4.5 Component for the Management of Time-Dependent Geologically Defined Geometries

A component for the management of time-dependent geometries should be able to manage complete time-stamped scenes -in the sense of the field-based approach- ("global time") and it should in the same way support an object-based time notion in which different times can be attached to the defined object states ("local time").

For a detailed digital modelling of geological processes like the development of a basin it is important that we can stop time for single objects and that we can go on with other objects. This is not possible with the scene-based notion of time. It allows the efficient access to an object set at a certain time, but most of the data are stored redundantly, also if only a small part of the data between two stored states of the database has changed. The scene-based time management stores snapshots of all non-temporal data after essential changes or in certain time intervals together with a time stamp. The implementation of such a data structure is straightforward: we only have to put the data of a data source regularly into an archive and to provide older versions together with the time of their back-up. A disadvantage of the scene-based time management is that the costs are very high to search for updates of single objects whose update time is not known. For the following example query a large part of the data has to be searched through and compared: "At which time the geological fault ROEVO of the Erft block in the Lower Rhine Basin was generated?" Queries that meet the snapshot of the whole data, however, are well supported in the scene-based time management. We give some examples:

1. "How was the appearance of the Lower Rhine Basin 1 million years ago?"
2. "How different is the recent geometry of the stratigraphic sequences in the Lower Rhine Basin compared to the geometry 1 million years ago?"

Conversely the object-based time management better supports queries about detailed changes at certain states like:

1. "How did the Erft block of the Lower Rhine Basin change during the last 1 million years?"
2. "What did the top of the main seam in the region of Bergheim look like at the time t_i?"
3. "Which fault did not change its position between the time steps t_i and t_{i+1}?"

As we have already discussed in chapter 4.2.3, both approaches of local and global time can directly be compared with the object-based and the field-based approach of geometric data (see chapter 2.1.2). In the object-based approach each object is coupled to a process. Thus time is a parameter of an object. Then at a certain stage static objects can be considered as states of an animation to be visualized. A method is sent to a dynamic process and this method returns a state. The sending and the update of the state can be taken over from the instance of a class that manages "time nodes". This central class manages besides a static object a corresponding dynamic process with local and present time. A process can be simulated by modifying the static object at a certain time. In the orthogonal concept of the global or scene-based time, however, for all objects of the scene and for each time cut a storage intensive snapshot of the whole scene has to be managed. Detailed sub-processes or single objects, however, cannot be described in this approach.

Geoscientific phenomena usually are described by objects and their relationships that both change in time. It is necessary to manage the time in an object-based way, because each single change with its corresponding temporal information has to be registered. It is not efficient to produce after each small change a new version of the whole non-temporal data as well. Furthermore, it is useful to store only those objects or parts of them between two times t_i and t_{i+1} (i ∈ 1..n) that have changed in this time interval. A spatio-temporal object then consists of snapshots for the object (parts) as its defined states.

We describe a temporal geometry object with the help of space-time complexes. Each space-time complex consists of a set of space-time simplexes that describe the geometry and topology of the object at a sequence of states. For each set of space-time simplexes *{STSimplex}* we define two mappings:

<u>pre</u>: *{STSimplex_{ti-1}}* → *{STSimplex_{ti}}*
<u>post</u>: *{STSimplex_{ti}}* → *{STSimplex_{ti+1}}*,

The *pre*-mapping returns to a set of STSimplexes at time t_i, the set of the STSimplexes at time t_{i-1} from which it originated, if it has been updated. The *post*-mapping returns to a set of STSimplexes at time t_i a set of STSimplexes at time t_{i+1} at which the set changes. If no update takes place, the input set of the *pre*-mapping and the output set of the *post*-mapping are empty. Thus the changing part of a space-time complex can be described with a set of STSimplexes whose elements are elements of the input set in the *pre*-mapping and of the output set in the *post*-mapping.

The set of all space-time complexes at a certain time (state) can be determined with an "event" to which a previous and a following event correspond. An event at time t_i then consists of a list of object changes at this state.

Changes in the geometric properties of geological objects usually take place during a time interval of many thousands of years. Exceptions are earthquakes, landslides etc.. It depends on the availability of the information and not on the relevant geometric shape of the objects, for which particular times the changes of the spatial properties are stored in the database as new states of the objects. The temporal distance between two directly following stored states usually differs. Not only the really stored time steps are of interest, but also the steps in between. Therefore it is necessary to enable an interpolation between the temporally neighboured objects. So the problem can be solved by storing only discrete states in the database, although geo-applications run in continuous time.

An interpolation between primitive simplex objects (simplexes) is straightforward. Complex objects are usually modelled as homogeneous collections of simplexes of the same dimension. In the simplest case the interpolation of complexes can be reduced to the simplex-to-simplex interpolation of the constituent simplexes. This simple case presupposes that 1:1 relationships can be constructed pairwise between all simplexes of both states. If an object has been approximated by a triangle network with k nodes, then all other object states also have to be approximated with k-nodes in each triangle network. This corresponds to the procedure in Worboy's time model (WORBOYS 1992) introduced in chapter 4.1.3.4. Obviously, this is not always the case. A geological object can change its size and/or shape in such a way that it needs more simplexes for an adequate representation than before. For example, in the result of deformations a flat platform (for the representation of which two triangles are sufficient) may transform into a spherical surface which will need a much larger number of triangles for the qualitative representation. In this case a simplex-to-simplex interpolation between two neighbour states is no longer possible.

We have already introduced the time modelling approach of POLTHIER and RUMPF (1995) in chapter 4.2.1.1. It has been realized in the GRAPE system (GRAPE 1997) and provides a simple interpolation between sequentially stored states in the database on the basis of an object--based time management. Precondition is that the geometric objects are represented as simplicial complexes. GeoDeform (ALMS et al. 1998), a GRAPE-based tool for the simulation of the geological restoration in the Lower Rhine Basin (ALMS et al. 1998) uses a 1:1 mapping of the simplexes for the interpolation between different states. The time model of GeoDeform works with the concept of the local and object-based time, respectively. In this approach a separated time can be attached to each animated object. The orthogonal concept is -as mentioned before- the global or scene-based time.

In the following we describe a component for the management of temporal geological objects on the GeoToolKit basis. It uses the time concept of GeoDeform and GRAPE for the visualization of the objects (Fig. 5.29). The *SpatialObject* class of GeoToolKit (BALOVNEV et al. 1997a) provides an interface for the handling of spatial objects. It seems reasonable to build the management of temporal object states on top of the spatial object class. The new class manages objects with spatial and temporal information, but without any thematic information. Each of these objects describes the development of the object's spatial extension in time.

We introduce the class *TimeStep* (Fig. 5.29) to represent different states of the same spatial object in time. This class contains a time attribute *(time)* and two references to spatial objects *(pre* and *post)*. They correspond to the representation of the objects at the time specified in the time attribute with its post- and pre-discretization factors. If the pre- and the post-discretization factor are equal, then the *pre* and *post* links refer to the same object. Geological strata that are constant in time can be modelled with GeoStore's *Stratum* class (see chapter 3.2.4). This class is a specialization of GeoToolKit's *Surface* class. So time-dependent strata can be modeled with the extended class *TimeStratum*.

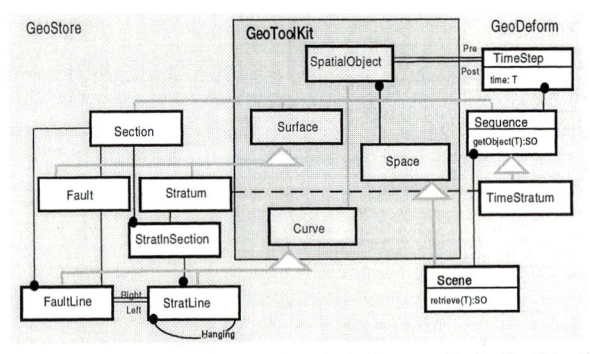

Fig. 5.29 Component for the management of temporal geological strata with GeoToolKit (from: BALOVNEV et al. 1998a)

A sequence of *TimeStep* instances that characterizes different states of the same spatial object is gathered into a spatio-temporal object (class *Sequence*). The TimeStep instances are ordered inside the time sequence which is given by their time attributes. Being a specialisation of the abstract class *SpatialObject*, a sequence can be treated in the same way as any other spatial object, i.e. it can be inserted into a space and participate in all geometric operations. The spatial functionality is delegated by default to the spatial object referred to in the latest *TimeStep* instance. Time-dependent objects do not exist continuously: an object can appear at a certain time and disappear at another time. To model this phenomenon a time sequence contains at least two *TimeStep* instances. A new instance can only be inserted into a time sequence, if its predecessor and its successor have the same discretizations. If the discretization factor of the new object distinguishes from the post-discretization of the predecessor or from the pre-discretization of the successor, a re-triangulation has to be computed.

A spatial object representation for a specific time can be extracted from the sequence by using the *retrieve* member function of the scene. A time-dependent selection for the interpolation differs from a common selection with a specified key. If there is no object in the time sequence that hits exactly the time stamp *t* specified in the retrieve function, instead of NULL it returns a pair of neighbour time steps with time values t_1 and t_2, so that $t_1 < t < t_2$. The same is valid for the time interval. If the interval's margins do not exactly hit the time step instances in the sequence, the resulting set includes all time steps fitting the interval completely extended with the nearest ancestor of the time step with the highest time value.

In the following we give an example of a temporal (1) and a spatio-temporal (2) query to be executed in the database component. Let *scene* be a container of type *Scene* and *bb* a pointer to a bounding box that specifies a three-dimensional region:

(1) *os_Set<TimeStep*>* tmp = scene.retrieve(time > -20.0 && time < 0");*

(2) *os_Set<TimeStep*>* tmp = scene.retrieve(bb, time > -20.0 && time < 0").*

In example (1) an object set is retrieved that existed from 20 million years ago until the present day. In example (2) additionally a spatial box is returned (bb stands for bounding box). Those objects are retrieved that existed at the given time interval inside this bounding box.

As *TimeSequence* is defined as a specialization of *SpatialObject*, any *TimeSequence* instance can be inserted into the *Space* class of GeoToolKit like every other spatial object and the retrieval is possible. However, to perform a temporal retrieval a special container class (*Scene*) is introduced which is capable of both spatial and temporal retrieval. This class is currently implemented with two separate components. The spatial functionality is completely delegated to the spatial sub-component. The temporal sub-component is implemented as a container which is indexed with the time attribute. ObjectStores functionality can be exploited to apply path indexes on set-based data members. The temporal retrieval can efficiently be executed by ObjectStore standard queries.

Figures 5.30 and 5.31 show the application of the time concept introduced above with time-dependent stratigraphic boundary surfaces and faults of the Lower Rhine Basin. They are to be seen as concrete examples of the change of the geometry and topology, respectively for geologically defined geometries at certain geological times.

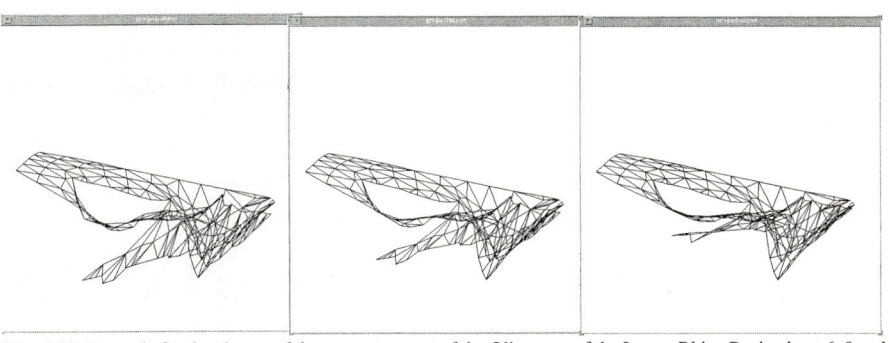

Fig. 5.30 Example for the change of the geometry: part of the Oligocene of the Lower Rhine Basin about 6, 9 and 13 million years ago[1], visualized with GRAPE

---

[1]. Figs. 5.30 and 5.31 are taken from: (BALOVNEV and BREUNIG 1997).

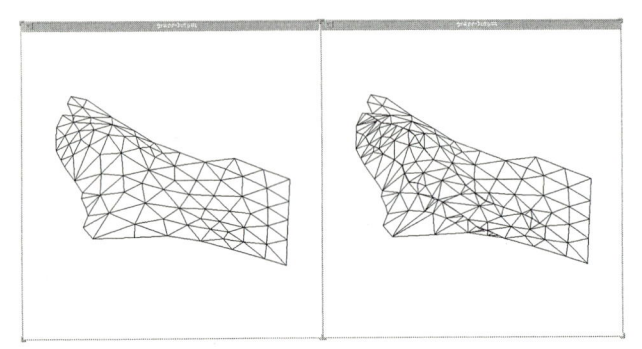

Fig. 5.31 Example for the change of the topology (discretization): part of the Oligocene of the Lower Rhine Basin about 28 million years ago, visualized with GRAPE

The management of temporal strata and faults is enabled with the additional attribute *"time"* in the corresponding objects. The respective attribute value shows at which time or time interval the object existed in geological "reality".

The database implementation of the proposed model, however, can result in a storage explosion, because the data are stored multiply. If, for example, only a small part of a large object changes in time, it is not useful to reproduce the stable part of the object in each time step. More promising is to subdivide the object into separate partitions and to consider only those parts of the geometry that are changing in time. GeoToolKit has to provide operations that can compose an object into a "static" and a "dynamic" part. A number of new spatial and temporal database queries can be applied as they are also interesting for multimedia documents (especially for simulations and video films). We give some examples of useful database queries for geological application:

1. "Return the geometry of the stratum body "main seam" 1 million years ago in the region of Bergheim."
2. "At which time did the strata A and B -which are separated by the fault S- have the largest distance between them?"
3. "In which time interval the strata A and B have touched each other along the fault?"
4. "In which time intervals did the main seam only consist of one connected body?"
5. "Return the change of volume for the main seam from 5 million years ago until today".
6. "Show the intersection of the Recent main seam with the main seam which existed one million years ago"

These database queries return a list of the stored object states during the time intervals specified in the queries. For the efficient realization of such queries an index has to be used for the management of time intervals (see chapter 4.3.3).

Chapter 6

Data and Methods Integration

One of the most important geospecific components for geoinformation systems in geology is the management of well data. In this chapter we show as an example a case study of well data from the Lower Rhine Basin, how useful it is for the geosciences to integrate heterogeneous data sources. The significance of the integration of heterogeneous spatial representations is again shown by an application in the North German Basin, Lower Saxony. Finally a data model integration is developed. It leads to the "intersection" of the geological with the geophysical 3D model which is presented in a specialized geophysical 3D modelling tool. In the following we distinguish between the meta data approach and the original data approach for data integration.

6.1 Starting with the Geoscientific Phenomenon: the Meta Data Approach

Going into the meta data approach we see that the geoscientific phenomenon, e.g. a geological process like the development of a basin, is the starting point of the data and methods integration. Thus the integration occurs at the meta data level. The first step is to determine which data and methods of different data sources are of interest for special geoscientific phenomena. They can either refer to similar data and methods of different examination areas or to data that have been captured with the same method in a common examination area. The database then serves as:

1. The retrieval of the locations for interesting data and methods;
2. The retrieval of meta data and methods;
3. The formulation of similarity queries;
4. The (graphical) output of selected data.

Seen from a systems point of view the meta data approach leads to a *retrieval component* for geoscientific data which can, e.g., be accessed in the internet. This component opens new data for the geoscientist to solve a certain problem. At the level of the meta data retrieval a statistical evaluation or a similarity analysis can be executed in a higher-level query module. Finally the integrated data is used for further interpretation of the data.

6.1.1 Retrieval of Meta Data

Important meta data occurring in many geoscientific data sources are the author, the captured date of the data and the method of the data acquisition etc.

Typical database queries for these meta data are:

1. "Which data have been captured from author A during the time from t_0 to t_1 by the method M?"
2. "Fade out the data of author A"
3. "How many % of the data have been captured during the time from t_0 to t_1?"

Depending on the application field statistical data also can enter into the meta data. In chapter 6.3.2 we will introduce a case study in which well data have been captured for the Lower Rhine Basin[1]. Typical database queries for meta data of drillings concern:

1. The drilling statistics (coordinates, bottom depth, surface height, etc.);
2. The header data (coordinates, drilling name, drilling number, drilling method etc.).

Spatial queries are of special interest to select the coordinates, the bottom depth or the surface height, i.e. queries to a set of drilliings in a one-, two- or three-dimensional interval. However, also other meta data like the drilling method used are of interest for the analysis of wells (BREUNIG and KLETT 1999).

6.1.2 Similarity Queries

It is extremely important for the geoscientist to get information at the meta data level about similar data or data that are comparable under certain aspects in other examination areas. "Similarity measures", however, cannot generally be defined, but they have to be determined in isolation for different application classes.

For the stratigraphic facies analysis, for example, which is based on results of geophysical drilling core measures (KLETT and SCHÄFER 1996) like interpreted measure curves, it is of interest to compare similar measure curves (see Fig. 6.3a). The similarity measure defined for use between electro log measure curves should compare the frequency spectrum and the occurring amplitudes by pattern recognition. However, hitherto this comparison has been done intuitively by the human eye, i.e. the similarity measure is influenced by many parameters of human expert knowledge.

6.2 Starting with the Data: the Original Data Approach

In the original data approach it is known a priori, which data are to be integrated. Ideally the data should be taken from one examination area and from one common context. The integration is directly executed with the original data. However, the meta data are also managed as

---

[1]. The data have been kindly provided by the Rheinbraun AG Cologne, Germany.

additional information and documentation of the data. The objective of this approach is to provide queries on top of database schemes that originated from different data sources managed by different software systems. The queries realized with "integrated data views" support the interpretation and the analysis of the data.

6.2.1 Integrated Database Views: Spatial Operators for Schema Integration

The general integration of heterogeneous (geoscientific) data sources is an unsolved problem, because semantic and technical problems avoid a general solution. Many authors have tried to develop approaches for this problem (FONG and GOLDFINE 1989; ABEL and WILSON 1990; ACM 1990; IEEE 1991; SHEPHERD 1991; ZHOU and GARNER 1991; WORBOYS and DEEN 1991; BILL 1992; BREUNIG and PERKHOFF 1992; SCHEK and WOLF 1993; ABEL et al. 1994; CREMERS et al. 1994; WIDOM 1995; LOMET and WIDOM 1995; BREUNIG 1996; BISHR 1997; BRANKI and DEFUDE 1998; DEVOGELE et al. 1998; BREUNIG 1999; etc.). Compared with usual business data the integration of geoscientific data leads to the following additional problems:

1. Different methods of data capture;
2. Different local meanings of the data;
3. The scale problem in space and time;
4. Heterogeneous spatial representations;
5. Different interpretations of geoscientific notions in the different disciplines.

All of the five listed points restrict the data integration significantly. With a pessimistic point of view one could argue that an extensive data integration in the geosciences is an illusion. However, the objective should be to homogenize the data for the integration of a special application or for a small set of applications.

The integration often leads to additional costs, because the data have not been acquired for this "integrated use". Thus it should be required that the data to be integrated have been captured with the same method and by comparable authors. Furthermore, geoscientific data often have to be seen in their local context. A comparison of data that have been collected for different examination areas is extremely difficult. Traditionally, the consideration of multiple scales is an important problem in the geosciences. Often geoscientific processes cannot easily be transferred by a generalisation or refinement. This means that the computation of a unique scale and the application of the integrated data to geoscientific models is problematic. Furthermore, often the temporal resolution of the data -like the period in special measures- has to be homogenized. The necessary interpolations, however, can lead to uncertainties.

Another difficulty that often is only seen as a technical problem is the modelling of the data in their different spatial representations. The conversion between the 3D representations leads to considerable problems concerning the exactness of the result (BREUNIG 1996). Often it is also not guaranteed that the back-conversion of spatial data leads to the original data which leads to inconsistent representations of the data.

Last, but not least, the integration of geoscientific data is difficult, because foreign data have to be included whose quality cannot be judged without an expert in the respective foreign field. This integration problem is relevant for many application classes like environmental monitoring, telecommunication etc.. Furthermore, the geoscientific notions are used with different meanings in the different disciplines. For example, the notions "model", "time series" etc. are differently used so that the integration of the data in a geoscientific model or within a time series needs a semantic structuring of these notions. Such a unification can only be achieved by many discussions between participating scientists. They could, however, lead to a buoyancy for the formalization of geoscientific phenomena.

A GIS can automatically solve two integration problems: first the problem of the homogeneity of different scales and temporal resolutions and secondly the problem of heterogeneous spatial representations. All the same the geoscientist has to weigh if such a homogeneity is useful or practical. Following BRANKI and DEFUDE (1998) we propose the following *spatial and temporal operators for the schema integration:*

1. spatialObject   changeSpatialResolution (resolution1, resolution2)

// changes the spatial resolution resolution1 of *this* and returns a spatial object with the spatial resolution resolution2.

2. temporalObject   changeTemporalResolution (resolution1, resolution2)

// changes the temporal resolution resolution1 of *this* and returns a temporal object with the temporal resolution resolution2.

3. spatialObject   changeSpatialRepresentation (representation1, representation2)

// changes the spatial representation representation1 of *this* and returns a spatial object with the spatial representation representation2.

We give three examples:

1. spatialObject   changeSpatialResolution (1 : 25.000, 1 : 50.000);
2. temporalObject   changeTemporalResolution (10.000, 100.000);
3. spatialObject   changeSpatialRepresentation (B-Rep, Simplicial3Complex).

The first example is referring to the scale of a geological map. In the second example the time scale is changed from 10,000 to 100,000 years, as this can be useful for the analysis of a geological restoration (see chapter 5.3). Finally in the last example the spatial representation of an object in the boundary representation is converted into a simplicial 3-complex, i.e. into a tetrahedron network.

A generalization has to take place to change the spatial resolution from a finer to a coarser resolution. The conversion in the opposite direction, however, is not possible without the processing of additional knowledge. Thus it often leads to an information loss. In BREUNIG (1996) implementation proposals are given for conversions as they are needed for the changeSpatialRepresentation operator.

6.2.2 A Case Study

Although there are serious problems of data integration it is possible to achieve success for certain applications. We demonstrate with the example of geological data in the Lower Rhine Basin, Germany how heterogeneous geodata can be modelled in different views of an object model. In the first step we consider the integration of different kinds of well data. In the second step we introduce the common management of these data with geological 3D model data to improve a geological 3D model of the sub-surface in the examination area.

6.2.2.1 Integration of Different Types of Well Data

In the Lower Rhine Basin there are suitable preconditions to examine, for example, the Tertiary sedimentary infill (SCHÄFER 1994) in detail. The correlation of the many wells requires the a priori analysis of all well informations. The well data usually contains the strata description. Besides the general information about the well like its name and its coordinates etc. the strata description contains for each stratum detailed information about the topographic map sheet, the archive number and additional descriptions like the stratigraphy, main lithology, accessory lithology, colour, palynological fossil content etc. and a set of well measurements. With the help of geophysical measurements, parameters like the gamma ray activity, bulk density, conductivity and resistivity can be determined. Such information can be gained from the strata descriptions or from the geophysical log curves. The integrated management of these data increases the chances for a detailed analysis significantly (BREUNIG and KLETT 1999). In the following we describe the two essential steps towards the incorporation of different types of well data:

1. Definition of an object model ("integrated views") for well data;
2. Definition of database queries for different types of well data.

Let us enter into the problems during the integration of the well data. The development of an object model - as shown here realized with the Object Modelling Technique (RUMBAUGH et al. 1991) - usually is a process of weeks or even months in which geoscientists and computer scientists specify the most important classes and their relationships in the framework of a requirement analysis. It is useful to define all important potentially usable database queries in parallel, so as to avoid or at least to reduce the probability that the object model has to be modified a posteriori. This leads to an efficient database design.

In a detailed analysis of the application three main classes have been identified: *lithological, geophysical* and *sample data* (Fig. 6.1). As a second step the relationships between these classes have to be specified. There are mainly hierarchical relationships (super-class/sub--class) in between lithological, geophysical and sample data. Corresponding to the main classes the object model can be subdivided into three views. Besides the lithological and the geophysical data the fossil data that have been collected as sample data are of special interest, because they describe the fauna and flora in their temporal development as biostratigraphic markers and are used to calculate derived climatic and ecological data/parameters. The specification of potential database queries and the incremental development of the well data classes with their relationships made it possible to define an object model that suitably maps the corresponding part of the "real world".

The most important implicit relationship between the three main classes of lithological, geophysical and sample data is the end depth of the wells, i.e. the z-coordinate in which the lithological, geophysical and sample data are located. For each predefined depth a description of a stratum object, a geophysical object and/or a sample object can exist. Finally, the administration module (user management) coordinates and controls the access to the well data and to the interpreted data (Fig. 6.1).

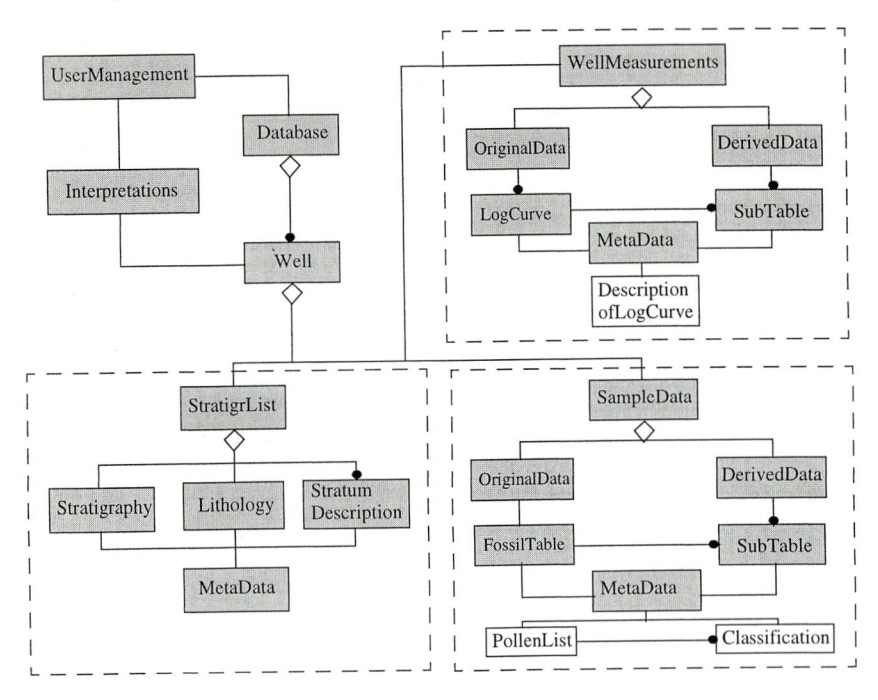

Fig. 6.1 Object model for well data of the Lower Rhine Basin, subdivided into three views

The detailed description of the specialized object models for the three types of well data is presented in (BREUNIG and KLETT 1999). We now enter into the queries for the original data which correspond directly to the three types of well data:

1. Strata descriptions;
2. Geophysical properties;
3. Sample data.

The *strata descriptions* provide information about the stratigraphy and the lithology. Against that the *well measurements* are gained with log tables. Furthermore, so called log curves are stored that can be used to control the interpreted data. The *sample data* are put together by *original data* and derivations *(sub-table)*. Especially climatic data are used for each pollen entry as *meta data* for the so called *pollen list*. Anticipating the integration of the different types of well data we distinguish between the following types of database queries for original data:

Chapter 6. Data and Methods Integration 149

1. Queries to common attributes for a set of wells

It is very useful for geologists to collect sets of well data which are selected according to special criteria in a so called "project". In this way user-defined well data sets can be specified and database queries to single attributes or to the combination of attributes can be defined. Examples for selection queries to attributes meet:

1. The well number;
2. The coordinates (spatial region query);
3. The end depth of the well;
4. The stratigraphic type (for example, according to the Schneider/Thiele scale and the biostratigraphy, respectively);
5. The stratigraphic values (e.g. only wells for which data about a certain lithology are existing).

The attributes mentioned above mainly refer to strata descriptions. We now enter into queries to single wells for different types of well data.

2. Queries to single selected wells

Queries to single selected wells are very useful for the further analysis of well data sets. These queries meet:

1. Measuring sizes of a depth interval;
2. *For the stratum description (SD):*
   - SD-entries (strata, lithological units inside a depth region);
   - SD-entries of a specific set of SD-entries (petrography descriptions of a stratum);
3. *For geophysical data sets:*
   - selection of two log measure values that are visualized as a plot of a statistical query result;
   - selection to the attribute "measuring size";
4. *For sample data of a well:*
   - frequency queries (e.g. "How frequent is the taxon *t* in all samples of the well *W*?");
5. *For sample data sets:*
   - selection of the sample type (group) and the depth (returns a corresponding measuring-value);
   - selection of the sample id (returns a corresponding measuring value);
   - selection of the climatic data of a well (have been collected for each pollen entry);
   - selection of the average temperature of the year, the coldest and the warmest month (integrated climate algorithm).

The query types mentioned above show the spectrum of possible interpretations for the data. The climate queries, for example, are very useful for the reconstruction of the climate conditions in the Lower Rhine Basin, Germany, during special periods of the Earth's history

(ARCHONTAS 1997). The plant fossils have been climatically evaluated with the so called coexistence method (UTESCHER and MOSBRUGGER 1995). From that a value for the corresponding climate parameter like the average temperature of the year or the sediment of a year is computed. The collected sample data, for example, give information about the composition of the vegetation in the Lower Rhine Basin for the deposition time of a stratum that has been sampled. The existence of different kinds of plants in a region allows conclusions for the regional climate. The coexistence method starts from the assumption that the climatic conditions most likely for the fossil plants are those which apply for comparable, vegetation communities existing today. The results of the coexistence method have been checked by Monte Carlo simulations of recent flora and confirmed by recent weather observations (UTESCHER and MOSBRUGGER 1995).

Obviously the samples taken correspond to certain geological strata which refer to a geological age determined by the geologist. The climatic database queries and the plots of log parameter values mentioned above are examples of advanced statistical queries that cannot be answered by standard database systems. An example is the database query for the resistivity in correlation with natural radioactivity which can be visualized in a plot. High values for the gamma rays are indicative for marine clays and low values for fluvial sandstones.

3. Integrated Queries

The real data integration cannot only be seen as the parallel processing of several types of well data. An important point is to provide so called integrated queries, i.e. such database queries as are using different data sources at the same time. We give some examples of such integrated and type-overlapping queries:

1. Combined queries for strata descriptions and log tables of a well:
   - selection of all attributes of stratum description with parameters of the log;
   - selection of log elements that correspond to a certain stratum;
   - selection of wells for which only geophysical data and not strata descriptions or sample data are provided.
2. Combined queries for sample data and stratum descriptions of a well:
   - "how often the taxa group T occurs in all clays of the well W?"
   - "how often the taxon t occurs in all yellow fine sands of the formation F?"

Figure 6.2 graphically shows the result of a combined query for a stratum description with different log tables. The geologist can draw further conclusions from the patterns of the curves and their similarities. For example, these conclusions can be used for the facies interpretation of the Tertiary in the Lower Rhine Basin. In the WellStore prototype system (BREUNIG and KLETT 1999) new interpretations of the geologist can also be stored in the database by editing the strata descriptions.

Chapter 6. Data and Methods Integration 151

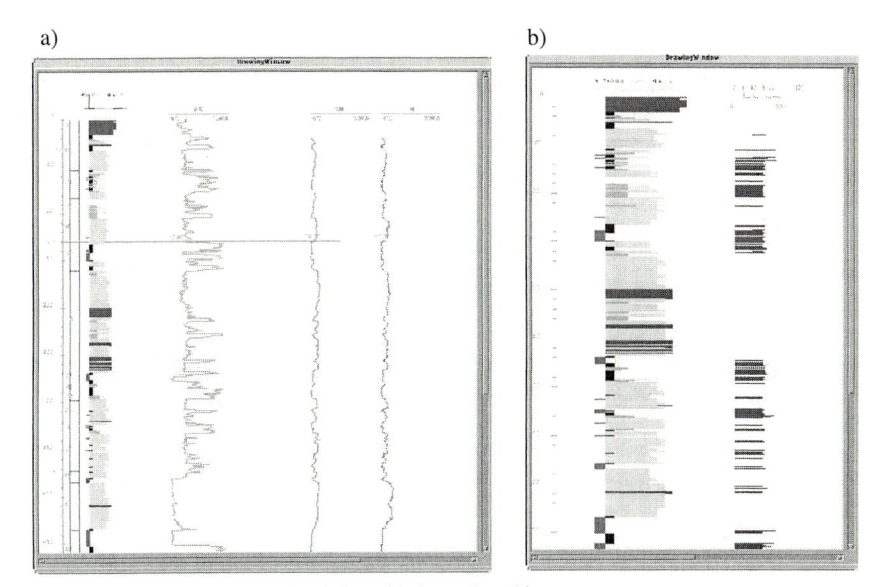

Fig. 6.2 a) Selection of a stratum description with three selected log curves
b) Selection of a stratum description with climatic values that have been derived by sample data

Figure 6.2b presents the result of a combined query from a stratum description and sample data. The left column shows the stratum description, while the right column presents climatic data that are determined by the climate algorithm mentioned above which has been integrated into the well database system. The climatic data correspond to the single strata of the stratum description.

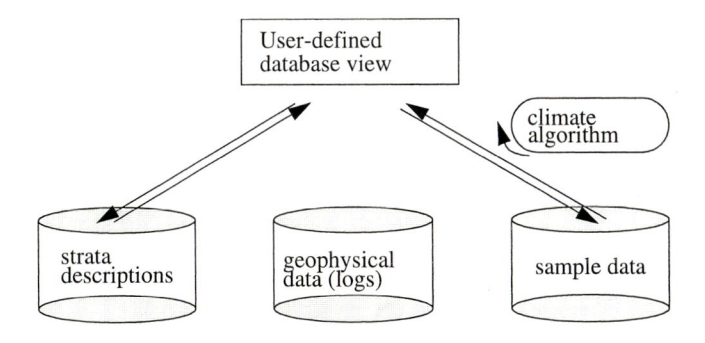

Fig. 6.3 Schema of the integrated query of Fig. 6.2 b

Figure 6.3 shows the schema of the integrated query from Fig. 6.2b for strata descriptions and sample data with the climate algorithm which is used by the database as an external service.

Problems caused by semantic heterogeneity of the data

The well data mentioned are coded with the so called "symbol key of geology" (PREUSS et al. 1991). In the course of time, however, many "dialects" of this coding key have come up. They especially contain additional commentaries of different geological applications. It cannot be the task of the computer scientists to solve such inconsistencies, but the involved geoscientists together are called upon to find a solution. In our case entries of the data which have not been correct according to the grammar of the symbol key of geology have been sent back to the geologists and corrected there. This pragmatic approach has resulted in a fast and efficient use of a unique format for well data.

Other problems of data integration that are closely connected with the process of data collection are the authorship problem and the problem of the reliability and exactness of the data. The first problem contains the juridical problem of the copyright of the data, especially if the data are used in different contexts. The quick interpretation of foreign data can lead to wrong results published by people who are experts in their research field, but not for the data used. To guarantee reliable data it is very important to provide a detailed documentation of data which should include the name of the author and the method of how the data have been collected. The user of any data should have the chance to fade out such data as he does not have confidence in. The case is even more difficult if data have been collected across different research fields. Then their reliability can no longer be judged objectively.

6.2.2.2 Integration of the Well Data with 3D Model Data

A qualitative improvement of 3D models -in our case the geological 3D model of the Lower Rhine Basin, Germany- can be achieved if we integrate the 3D model data (of the application in chapter 5.3) with the well data introduced above. In the application, already interpreted sections served as "original data" for the construction of stratigraphic boundary and fault surfaces. The wells can be used as additional constraints for the triangulation of the surface data. This leads to a refinement of the 3D model. Additionally, a correlation of the wells can be aspired to. The correlation has to be supported by a well data editor which also allows to store new interpretations for lithological sections. A prototype of such a well data editor has been developed for well data of the Lower Rhine Basin (BALOVNEV et al. 1998a; BREUNIG and KLETT 1999).

From a systems point of view this means that the well data component has to be integrated into a system for the management of geologically defined geometries like GeoStore (BODE et al. 1994). The integration enables to formulate "integrated queries" for 3D model data and well data at the same time. The geometry of the wells -which in a pure well data component is not explicitly needed- can be described as a segment in three-dimensional space implemented in a geodatabase kernel system. In our case this means that the GeoToolKit class *Segment* can be used from the well data component. Information about wells, sections and faults can then be taken from the database browser (Fig. 6.4). The other way round selected sections and faults can be visualized in 2D and 3D space.

Chapter 6. Data and Methods Integration 153

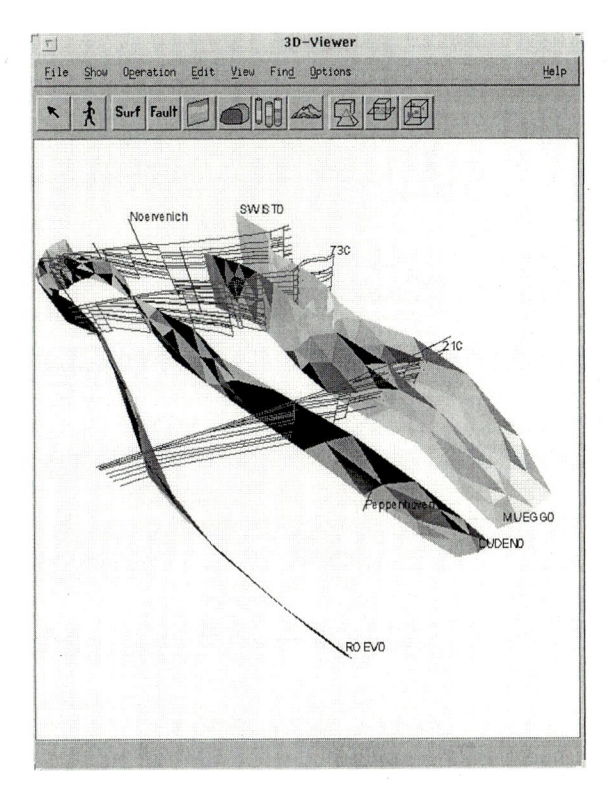

Fig. 6.4 Integration of well data with digitalized sections and 3D model data

It is very useful to insert the well information into the 3D model. The lithology of the wells can directly be compared with the geometries of the single stratigraphic boundaries and stratum bodies, respectively.

6.3 Integration of Heterogeneous Spatial Representations

6.3.1 The Problem

The high significance of the geometry and topology for geoscientific data implies that the different geometric and topological representations -in the following named as spatial representations- are central for the integration of heterogeneous data in the geosciences (BREUNIG 1996). In the following we consider examples for 3D representations as they are used in geoscientific software tools.

6.3.2 Spatial Representations Used in Geoscientific 3D-Tools

Depending on the application spatial representations are differently optimized in today's geoscientific software tools. For example, the representations of solids in the 3D modelling tool GOCAD (MALLET 1992a) is optimized according to an efficient visualization of the objects. GOCAD is mainly used for geological and medical applications. In GOCAD solids are modelled as sets of tetrahedra, the so called *Tsolids*. They are implemented as TSOLID container classes which are derived from the simplicial complex classes. The data members like vertices and the corresponding tetrahedra are derived from the corresponding atomic geometry classes.

To take another example, in the geophysical 3D modelling tool IGMAS (GÖTZE and LAHMEYER 1988) solids are optimized according to the efficient computing of potential fields. In IGMAS solids are represented as hulls, i.e. as closed surfaces. Each hull is an approximation of a surface which is implemented as a triangle network. Furthermore, to each triangle network 2 or in the general case 2*n attributes (n>0) are attached. The interior and the exterior density of a solid are an example of such a double attribute. IGMAS contains a geophysical 3D computing component in which for example the computed and the measured density or the magnetic susceptibility of geological solids can be compared with each other.

Finally in GeoToolKit (BALOVNEV et al. 1997a) the developers payed attention to an efficient database realization of the geometries. GeoToolKit's abstract geometry classes allow the user to select between different spatial representations. The geologist, for example, can choose solids that are represented as tetrahedron networks (simplicial 3-complexes). The GeoToolKit class *gtTetraNet* contains the following data members:

1. An R*-Tree for the storage of the 3-simplexes;
2. A counter for the number of the 3-simplexes;
3. The bounding box of the simplicial 3-complex.

ObjectStore's functions for the allocation of persistent storage guarantee the clustered storage of the 3-simplexes of a simplicial 3-complex. The bounding box and the R*-Tree data members enable fast access to single 3-simplexes inside the tetrahedron network. In the class *gtTetraNetElement* each object has a pointer to its four neighbouring tetrahedra. The neighbours are explicitly used to optimize 3D-operations like the intersection of tetrahedron networks.

The integration of heterogeneous spatial representations for geo-objects can be realized via a conversion between the different spatial representations or via a transformation into a "neutral format". However, in principle information losses cannot be avoided. In BREUNIG (1996) and in BREUNIG (1999) we have already presented approaches for the integration of heterogeneous spatial representations which are based on convex simplicial complexes. In the following chapter we present a geological and a geophysical application which exchange their geo-objects among one another.

Chapter 6. Data and Methods Integration 155

6.4 An Example from Geology and Geophysics

6.4.1 Objectives of the "Integrated Application"

We introduce a geological and a geophysical application from Southern Lower Saxony, Germany (Fig. 6.5).

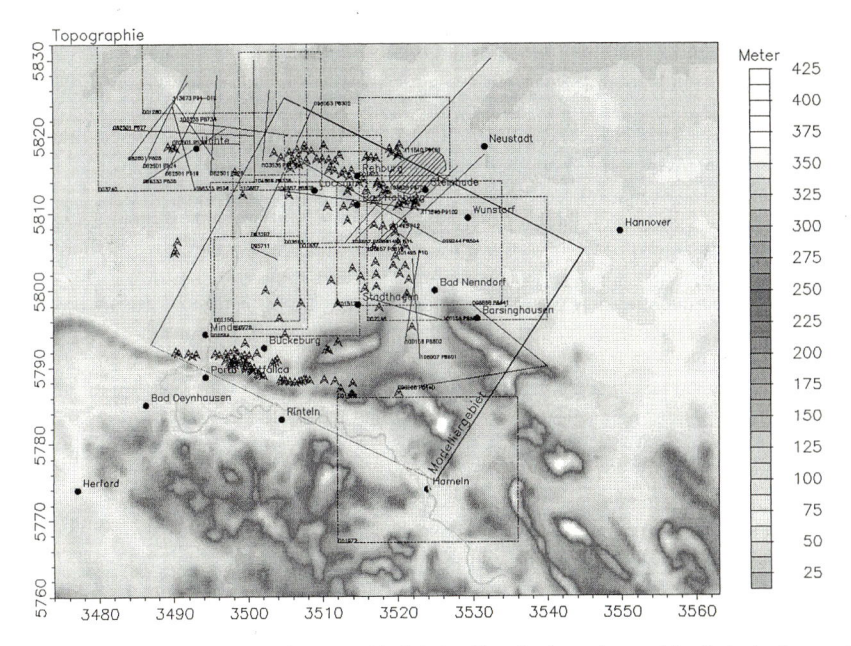

Fig. 6.5 Topography of the examination area with digital wells, seismic sections and detailed seismic examinations[1]

The central objective of this application is to construct a new type of geological map with the support of geological and geophysical 3D models (SIEHL 1988, 1993; BREUNIG et al. 1999). In the examination area there is still onging exploration for hydrocarbon deposits by different oil companies. This leads to a high density of geological and geophysical measurement data[2]. For geology the data sources are isoline plans and sections of the Geotectonic Atlas of Northwest Germany (KOCKEL et al. 1996). For geophysics the measured density and magnetic fields, information about the seismic properties and about the geophysics of the wells is of special interest. The goal is to construct a geological map as the intersection of the digital terrain model (DTM) with the 3D stratigrahic boundary surface model. Finally the geological model has to be evaluated with the 3D density and susceptibility model. In

---

[1] The figure has been generated by the group of HaJo Götze, Geophysical Institute of FU Berlin, Germany.
[2] The data have been kindly provided by the involved oil companies BEB Erdgas and Erdöl Hannover, MOBIL AG Celle, WINTERSHALL AG Kassel, PREUSSAG AG Lingen and the NLfB in Hannover for the IOGIS subproject "Geological Map" (Si 73/15-1).

total more than 1000 wells have been used, also more than a dozen digitalized isoline maps that are based upon the Geotectonic Atlas and a few vertical intersections. The isoline maps are based upon the evaluation of the reflection seismics by KOCKEL et al. (1996).

Together with the wells, the depth maps serve as the starting point for the geological-geophysical 3D modelling of the examination area.

Analogously to the interactive 3D modelling process described in chapter 5.4.3 sections, stratigraphic boundary and fault surfaces and finally stratigraphic solids can be constructed. The digitalized isoline maps serve as the basis for the construction of stratigraphic boundary surfaces and the wells are used as additional base points. In the examination area the average number of triangles per surface was 10,000. One of the difficulties was to intersect the stratigraphic surfaces with the DTM and with the faults.

The adaptation of the stratigraphic surfaces to the depth positions of the digitalized isolines can, for example, interactively be constructed with the Discrete Smooth Interpolation method (GOCAD 1996). All the boundary nodes of the stratigraphic surfaces have to be considered; i.e. the original nodes as well as the nodes that have been generated during the intersection operations have to be searched through. Furthermore, all the points of the digitalized isolines have to be taken into account as constraints for the interpolation. For the application described above 50 iteration steps have been necessary. This is a time-consuming process. The constructed stratigraphic surfaces can be applied in a first approximation for the horizons of a geological 3D start model (Fig. 6.6[1]).

This first 3D starting model has to be extended in a second step by the geophysicists with the definition of the density for the different geological units. For example, the model can be used as the basis for the 3D modelling of the density for the Earth´s surface at the base of the Zechstein (Upper Permian).

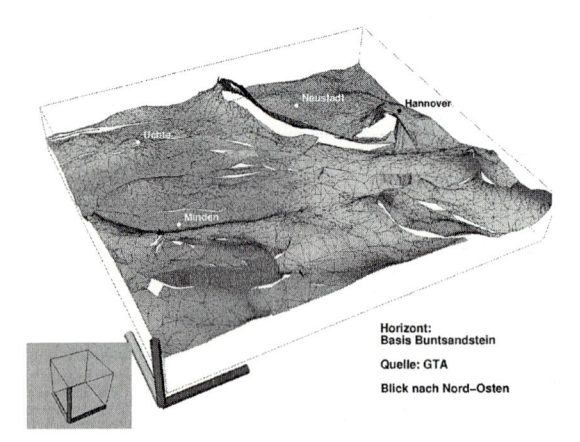

Fig. 6.6 Triangulation of the horizon at the base of the Bundsandstein (Lower Triassic, 12711 triangles)

[1.] Figure 6.6 has been generated by Christian Klesper, GFZ Potsdam with the Vis.tool IVIS-3D.

At the data modelling level, the bridge between the geological and the geophysical application has been built by the explicit 3D modelling of physical stratum properties and their geometries.

6.4.2 Consistent Object Exchange and Data Model Integration

The iterative 3D modelling process requires a cooperative work between the geological and the geophysical 3D modelling tools on the same data. Thus the first step is to bind the geological and the geophysical data in a common data model. There are general, but incomplete data model standards like OGIS (BUEHLER and MC KEE 1996) and POSC (1997) which follow up similar goals. However, their implementations are not yet sufficient for highly specialized applications. Concrete models that are used in different software tools have also to be considered. It is necessary to develop a flexible data model that provides different views upon geological and geophysical data. The Object Modelling Technique (RUMBAUGH et al. 1991) and the Unified Modelling Language (UML 1997) provide a suitable object-oriented method as a framework for the data model integration.

The integration of geological and geophysical data in a common data model requires the checking of the integrity and consistency of the "integrated" data model. The data model should avoid the storage of inconsistent data. To guarantee the consistency of the data we need semantic knowledge about the data types. Thus meta information should be available about the structure of the data types. The question as to which information is really necessary to hold the data consistently can only be answered after years of software development work. It is also not useful to define a model with a fixed type set. The model should allow the easy extension of new data types.

In many applications the attachment of thematical information to the geometry and topology of geo-objects is not unique and can vary with different applications. In geology and geophysics different ways can be shown regarding how the geometric and topological information can be coupled with the properties of the materials. We enter into different "space partitions" in geo-logy and geophysics, respectively.

1. "Stratigraphic Partition"

The superior space and time partition in geology today can be done by use of the chronostratigraphical scale, supported by the radiometric and bio-stratigraphical data of the strata. The resulting stratigraphy provides a time scale for the geological processes and events and is thus the basis for the reconstruction of the history of the Earth and life on Earth.

2. "Facies Partition"

Sediments in geology can also be classified by their lithofacies or biofacies which describes the rock type or the fossil contents of certain facies fossils, respectively. The partition into different facies leads to a completely different space partition. It is more oriented on the paleogeography and the paleoecology than the time partition of the stratigraphy. One stratum can contain several facies partitions.

(3) "Density Partition"

In geophysics among others the density inside and outside geological solids are of interest. The space can be subdivided into volumes of different density. We call this space partition the "density partition".

In principle the question arises, if for each attribute like "stratigraphy", "facies" and "density" its own space partition should be chosen. Alternatively, for each geometric primitive of an object -for example, a single point or tetrahedron- different attributes could be referred to that corresponding to the different space partitions. We choose the first possibility because it avoids data redundancy.

New conclusions can be drawn from interdisciplinary problems if, for example in geology and geophysics, the space partitions which have been developed by different geological and geophysical methods, are seen together. For this the "intersection" of the space partitions is necessary. We give an example for the integrated view of the stratigraphical, the facies and the density partition.

a)

stratigraphic Solid
stratigraphy

densitySolid
density

b)

faciesSolid
facies
stratigraphy

densitySolid
density

c)

SolidCertDensity
facies
stratigraphy

densitySolid
density

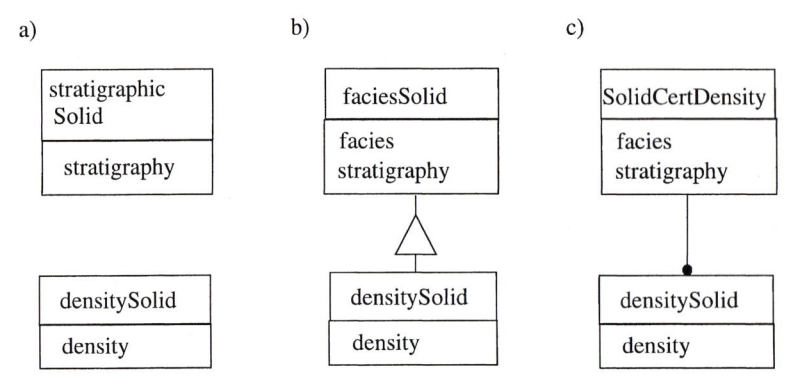

Fig. 6.7 Different object-oriented modelling of geological and geophysical solids
    a) Independent stratigraphy and density partition
    b) Density solid derived from a facies solid
    c) Solid of a certain density which is related to a density solid by an association

In Figure 6.7a each of the two classes *stratigraphicSolid* and *densitySolid* independently of each other has its own space partition for the stratigraphy and the density. Against that in Fig. 6.7b the class *densitySolid* inherits from a general class *faciesSolid*. This corresponds to the physical fact that the lithology and the often synonymously used stratigraphy determines the density of a solid. Figure 6.7c allows to subdivide the solid of a certain density *(SolidCertDensity)* into a set of density solids for the case that in a distinct facies or stratigraphy domain different density values occur. As in Fig. 6.7a both space partitions are independent of each other. That is why the geometries of the density solid -which corresponds to a stratigraphy solid- have explicitly to be computed. This means that the intersection of the two solids has to be determined. In Fig. 6.7b the intersection is already modelled "hard wired" in the object model.

The "intersection" of the different space partitions with the material properties can be executed analogously to the intersection of map layers in a traditional 2D geographical information system. The attributes are "unified", for example the intersection of a geological solid with the facies value "sandstone" and a density solid with the density value "2.7" results in a density with the union of these two attribute values. Analogously geological solids or geophysical solids could be intersected with their different attributes like the intersection of the density with the susceptibility. Geometrically the consequence is a finer space partition that results by the geometric intersection of both given space partitions.

Another interesting question is how the change of the material properties in time can be modelled. In the case shown (a) this is easily possible, because the stratigraphy and the density distribution are independent of each other. Thus they can also be versioned independently of each other. In the modelling cases (b) and (c), however, the change of the density inside a geological solid in time would result in a change of the topology and the geometry.

We now derive a geological-geophysical 3D object model that simplifies the object exchange between geological and geophysical applications[1]. As we shall see, the geological and the geophysical solids are derived from pure geometric classes.

6.4.3 Geological-Geophysical 3D Object Model

The mapping between heterogeneous spatial 3D representations is a central point for the integration of three-dimensional geological and geophysical objects. For the time being let us reduce the integration problem to this point. Then we can state:

The geophysicist is mainly interested in the hull of solids and in a few material properties like the interior and the exterior density of solids. For the geologist, however, especially the internal fabric of the solids is of interest as well as the spatial association of solids, their topological relationships and differentiated material properties.

A deformed geological stratum intersected by faults consists of a set of non-connecting or only partly connecting solids that are bounded by stratum and fault surfaces and have distinct lithological properties. The stratigraphic information of the geological stratum can be used to achieve an estimation of the density in the geophysical tool. For the geophysics tool -take as an example the IGMAS system (GÖTZE and LAHMEYER 1988)- it is important that the solids must not contain "holes" and especially the boundary surfaces have to be closed. Only in this case can the interior and exterior density of solids be computed.

The derived attributes *getSolid* and *getSurf* of the geometric kernel (Fig. 6.8) transform mutually solids and surfaces, respectively, in the geological and in the geophysical model. Attribute *getSolid* determines the solid from a given hull with a tetrahedralization and *getSurf* returns the hull which surrounds a solid as a closed triangle network. Thereby the object exchange is guaranteed in both directions -from geology to geophysics and the other way round.

---

1. The object model has been constructed in close cooperation between the geological group of Agemar Siehl, University of Bonn, the geophysical group of HaJo Götze and Sabine Schmidt, FU Berlin and the Computer Science group of Armin B. Cremers, University of Bonn.

160   *Chapter 6. Data and Methods Integration*

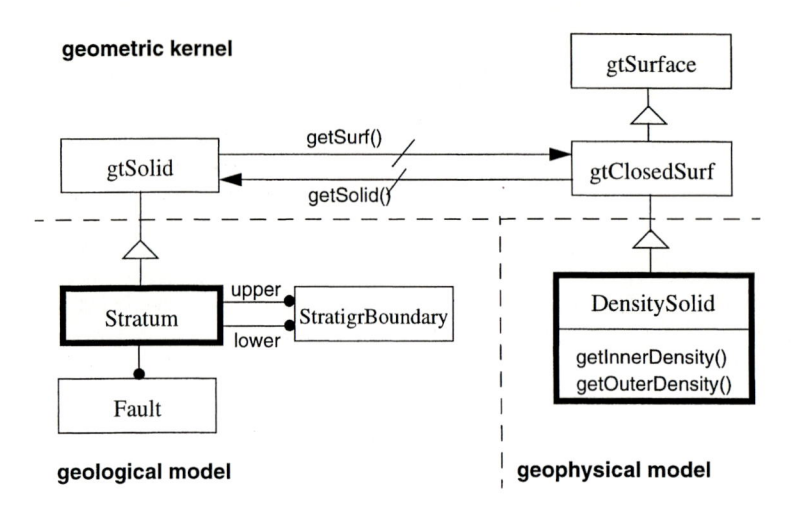

Fig. 6.8 Geological-geophysical 3D object model

The part of the geological-geophysical data model in Fig. 6.8 maps complex geological facts. The following geological situations can be modelled:

1. In the vertical direction a *Stratum* is bounded by an upper and a lower stratigraphic boundary *(StratigrBoundary)*;
2. In the horizontal direction a stratum is bounded by one or more *Faults*;
3. In the horizontal direction a stratum is bounded by the "model boundary"[1], or by reduction of its thickness to zero;
4. In the vertical direction a stratum is also bounded by stratigraphic boundaries respecting the Earth's surface.

Consequently the object model is more general than the model for the Lower Rhine Basin which has been introduced in chapter 5.4.1.

If we also consider the geological and geophysical meaning of the surfaces and solids, we have to develop an object model as presented in Fig. 6.9. It originated from a close cooperation between geologists, geophysicists and computer scientists. It shows the relationships between the geological and geophysical objects in more detail than the first object model of Fig. 6.8.

---

[1]. This case is not explicitly modeled in the object model, because the "model boundary" is already implicitly given with the bounding box of the geometries in the geometric kernel.

Chapter 6. Data and Methods Integration 161

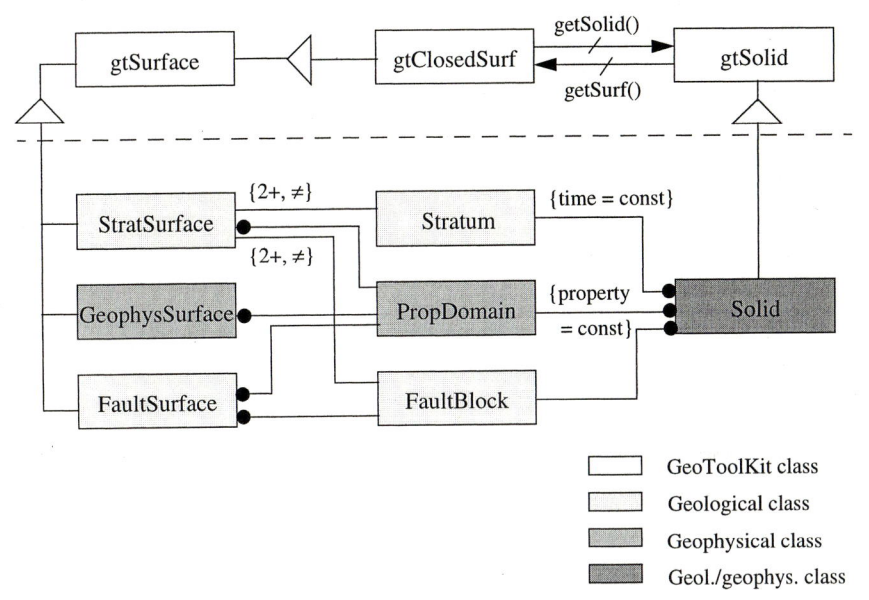

Fig. 6.9 Extended geological-geophysical object model

On top of the geometry classes of GeoToolKit differentiated geological and geophyscal classes are defined. The topology between the objects is also considered. For example, to one *Stratum* two stratigraphic boundaries are corresponding -an upper and a lower one. Each *FaultBlock* can have a set of fault surfaces. A geophysical domain with homogeneous attribute values *(PropDomain)* can be bounded by a set of stratigraphic surfaces *(StratSurface)* and itself consists of a set of surfaces *(GeophysSurface)*. A stratum with the same lithology, a fault solid and a geophysical domain with the same attribute values can consist of a set of *Solids*, respectively. The geological and geophysical surfaces *(StratSurface, PropSurface* and *FaultSurface)* and *Solids* inherit the geometric functionality of the GeoToolKit classes.

6.4.4 Methods Integration

Essentially for the practical work not only the integration of the data, but the "integrated use" of the methods is across systems boundaries. We will explore this point in chapter 7. In advance we give an example for the methods integration across the boundaries of the both 3D modelling tools GOCAD (MALLET 1992a) and IGMAS (GÖTZE and LAHMEYER 1988).

The interactive graphical and geophysical modelling tool IGMAS (Fig. 6.10) interprets the measured potential fields of the gravimetrics and/or of the magnetics with numerical 3D modelling computations (GÖTZE 1978, 1984). Thus the gravity which has been measured at the surface of the Earth is compared with that which has been modelled and computed for the sub-surface. With this comparison the geophysicist can improve the model of the density of the geological solids. The modelling is mainly executed as batch services, i.e. the implemented methods are invocated to compute the density on their own.

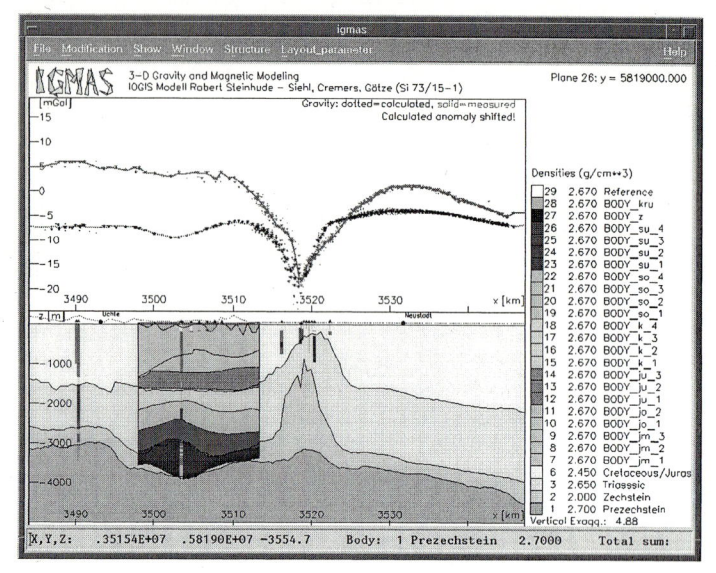

Fig. 6.10 IGMAS example of a section from the Northern German Basin in Lower Saxony with the density distribution and the overlapping of the geological 3D model that has been generated before in GOCAD[1]

The special merit of GOCAD (MALLET 1992a) is the possibility of a very flexible interactive 3D modelling of geological surfaces and solids. This is supported by an efficient 3D visualization and by interactively usable geometric interpolation algorithms.

The 3D model computations can start and the density of the solid can be computed as soon as IGMAS gets the hull (surface) of a solid that has been modelled in GOCAD via the GeoToolKit. The geological and the geophysical 3D model are completed by the GeoToolKit classes. With the coupling to GeoToolKit IGMAS can start from its own user interface database queries via browser to the database server. For example, the result of geometric 3D opeations of GeoToolKit can be transferred to IGMAS. In GOCAD the geophysically defined solids can then be visualized and compared with well data. That is how the geological 3D model can further be refined. Figure 6.10 shows the overlap of the geological 3D model with the density representation of IGMAS.

---

[1]. The figure has been made by Sabine Schmidt (group of HaJo Götze, geophysics FU Berlin, Germany). The geological 3D model has been developed by Robert Seidemann (group of Agemar Siehl, geology, University of Bonn).

Examples for typical geological, geophysical and combined geological-geophysical database queries are:

1. "Return the thickness and the volume of a stratum S inside the given boundaries";
2. "Return the intersection of a plane with a facies solid inclusively its attribute values";
3. "Return the stratigraphy solid inside the specified region";
4. "Return the geometry and the density of the solid D";
5. "Return the attribute values of the neighboured density solids";
6. "Return the boundary surfaces to the given density attribute values";
7. "Return the attribute intersection inclusive the geometry of all geological and geophysical solids".

In principle it would also be very interesting for geophysicists to follow the density values in time over the years ("4D modelling"). An extension of the 3D model in IGMAS for the different object states is possible. However, for all times that are older than 100 years there are no empirical values to compare. Thus the computed density cannot directly be evaluated. But as the exogene geological processes like erosion, transport and sedimentation are all controlled by gravity, paleo-geophysical models could thoroughly help to interpret paleogeographical scenarios.

In the next chapter we introduce an architecture for a component-based GIS. From a systems point of view the data and methods integration for the shown geological-geophysical application can be realized with this architecture.

Chapter 7

Systems Integration

After a short introduction into the requirements and the approaches to systems integration and the coupling of geoinformation systems, we again enter into the geological-geophysical application of chapter 6. We present the implementation of a component-based GIS, which has been realized distributed in Bonn and Berlin, respectively and is based on a CORBA systems architecture. The component-based GIS consists of a geological, a geophysical and a database server component. It is shown that geology as well as geophysics profit from the coupling of the geological with the geophysical component via an open geodatabase.

7.1 Requirements

The coupling of hitherto isolated information systems and of modelling tools in the geosciences allow an interdisciplinary synthesis and a digital checking of models and analysis methods (OGIS et al.1997). In this chapter will we show that synergy effects in the geosciences can be achieved with the development of a *new generation of component-based geoinformation systems*.

One of the most important requirements from the geosciences is the distributed use of methods in distributed GIS components. The kernel of the efforts is to achieve "interoparability" between geoscientific software components. During the systems development the adaptibility and extensibility has to be guaranteed. With interoperability we mean in this context the communication between distributed software components to solve common geoscientific problems. Today's software systems like databases and geoinformation systems, however, often have been developed isolated from each other. Each software component provides a reduced functionality so that the components can effectively interoperate with each other (CREMERS et al. 1992). Existing integration approaches mostly are realized as closed couplings and they are not compatible with component technology.

The architecture of geoinformation systems has to be changed so that functionality will be distributed to single components and special services have to be provided for foreign systems. Each component of a distributed GIS should encapsulate its implementation and interact with its environment with uniquely defined interfaces. This increases the degree of reusability. For example, a service for external data is useless for the geoscientist if the location, the author and the methods for the data capture are not described in detail. The geoscientist calls this information "meta data" as already mentioned in chapter 6. In contrast to the notion of meta data used in computer science, meta data in the geosciences do not correspond to the structure of data and methods, but exclusively to their contents. The geoscientist often also speaks of "header data", because they usually are described at the heads of files.

Following the component notion of SZYPERSKI (1998) we also allow instances of components that have a persistent state like database server components. Thus copies of component instances have to be distinguished from the components themselves.

7.2 Approaches for Systems Integration

During the last twenty years the requirement of data exchange has been one of the most exciting challenges for the development of geoinformation systems. In earlier times the data exchange mostly has been realized with a file interface or with coupling programmes that have been closely coupled with GIS. This approach implies that $O(n^2)$ different coupling programmes had to be written for n interfaces (Fig. 7.1a). The realization of a common interface -preferably an object-oriented solution- only needs $O(n)$ interfaces (Fig. 7.1b).

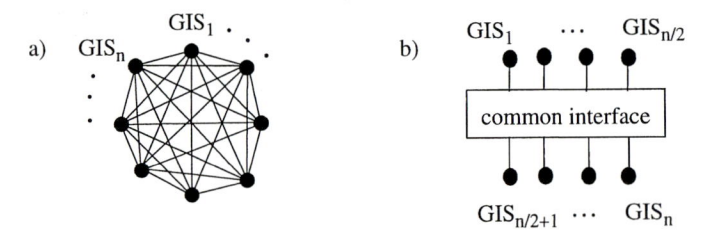

Fig. 7.1 Architecture of the systems integration with single coupling programmes and a common interface

The integration between two GIS A and B can be realized by the following approaches:

1. Data exchange and file transfer between the software systems A and B, i.e. realized with a converter between the geodata formats of the systems A and B.
2. Inclusion of the functionality from system A into system B or vice versa.
3. Data and methods exchange with remote access.
4. Data and methods exchange with the local mirroring of data and methods of the remote application (local access).
5. Combination of approach (3) and (4).

Approach (1) can be seen as the traditional approach of systems integration (ABEL and WILSON 1990). The data exchange is realized by export and import interfaces. The methods of software tools, however, cannot be used remotely with this approach.

Approach (2) can easily be realized. The former functionality of system A is re-implemented in system B or vice versa. Conversion problems of the data can be solved internally in system B and system A, respectively. This approach, however, leads to complex system architectures and avoids the modular design of the system. This often leads to a cumbersome control of the system. An example of such a solution is the integration of a geophysical seismics component into the 3D modelling tool GOCAD (MC GAUGHEY 1997).

Approach (3) guarantees the modular software architecture of GIS (VOISARD and SCHWEPPE 1994; BALOVNEV et al. 1997b). It can be well used in local area networks, but can also lead to a poor performance in wide area networks. In the following we call this solution the "local approach".

In the local approach an application has different programme parts that are distributed over the network. It copies the source code of these components to its own client. Java beans, for example, provide platform independent code that can also easily be integrated into world wide web protocols. The so called applets are objects that are instantiated by a special java class. This class and other interfaces allow the applets to be executable in web browsers. User-defined applets can extend the standard class. Inheritance enables the applet to be the basis of a graphical user interface (GUI). In this approach the code for the class which provides the methods to be called by the user are transferred over the network; i.e. the method is not executed remotely, but the applications are locally run on the client. The result is that the methods are executed on local object instances.

Advantages of the local approach are:

1. The distributed application can completely be executed on the local client. This increases the performance significantly;
2. Internet programming is directly supported.

Disadvantages concern the following aspects:

1. The programming of distributed applications is not supported;
2. Most of today´s GIS applications are not written in Java.

There are many examples of geoinformation systems that are using Java applets (KOSCHEL et al. 1996; FITZKE et al. 1997; etc.) and their number is increasing rapidly. However, the author is not aware of complex distributed GIS applications of this kind.

Approach (4) avoids the disadvantage of approach (3), but the programme code has to be copied over the network from one computer to another. Thus inconsistencies can occur between the original version and the copied one. In the following we call this solution the "global approach".

The "global approach" provides a systems architecture that allows to call operations of objects arbitrarily distributed in the network. They are used in the same way as the local objects of the application. The Common Object Request Broker Architecture (CORBA) provides an object-oriented standard infrastructure for the access on distributed software components in heterogeneous networks (MOWBRAY and ZAHAVI 1995). The components can be remotely used, i.e. their source code does not exist on the local client computer.

Important *Advantages* of the global approach are:

1. Object-based data and methods exchange over arbitrary hardware platforms and operation systems to objects of heterogeneous programming languages;
2. Direct support of the programming for distributed applications;
3. Use of standardized basis services and of software that is already available in the network.

Obvious *disadvantages* of the global approach are:

1. Computation intensive and I/O intensive applications are dependent on the performance of the network;
2. Internet capable programming and the embedding into web browsers is not directly supported.

KOSCHEL et al. (1996) compare different CORBA implementations for use in geoinformation systems. We propose to combine approach (3) with approach (4) into a new approach (5). A precondition for this new approach is a software architecture that is already subdivided into components. CORBA and Java technologies can be combined, if a mapping from CORBA's Interface Definition Language (OMG IDL) to the Java programming language is defined. The result then is a Java Object Request Broker (Java ORB) combined with a runtime system that supports the language mapping.

The CORBA user gains with the Java embedding of the OMG IDL the portability over platforms and the chance to use internet programming (VOGEL and DUDDY 1997). The Java linking allows the implementation of CORBA clients as applets. This enables the access to CORBA objects over the web browser.

The Java user also gains from the use of CORBA: Java does not support the direct development of distributed applications. The Java linking for OMG IDL, however, provides the application programmer with the distributed object paradigm of CORBA. Interfaces are provided that are independent of the implementation. The access to objects in other programming languages and to objects anywhere in the network is possible and the automatic code generation for remote access is enabled. Furthermore, the CORBA services and facilities can be used.

The combination of both approaches seems to be particularly well suited for geoinformation systems, because there are computation intensive specialized components as well as a large spectrum of data and methods components. The GIS user has access to both types of components.

It is useful to execute the time critical GIS operations locally and to outsource standard services into distributed components at different locations in the network. This solution also includes the extensive use of already existing GIS software like 3D visualization tools.

7.3 Coupling Mechanisms

In the following paragraphs we orientate ourselves to the statements in BALOVNEV et al. (1998b) and (BREUNIG et al. 1999).

7.3.1 Client-Server Connection of an OODBMS

The concurrency problem of distributed GIS can partially be solved, if the client components access the concurrency control mechanisms of a database server. The architecture of modern object-oriented storage systems like ObjectStore, however, require that the complete data processing including database queries is executed on the client side. This principal design solution has certain advantages in respect of the efficient maintenance of persistent objects. However, it can also lead to an unjustifiable overload of the network. For example, when we need to pick a single object from a large collection, the whole collection has to be transmitted through the network from the server to the client in order to test the selection predicate. A network that is not fast enough can become a bottleneck in the performance. The alternative solution is to perform the necessary selection on the server side and to transfer only the selected object to the client. That means practically, we should ship an operation through the network which would be applied to the corresponding data on the server side.

7.3.2 Client-Server Connection via Standard UNIX Technology

Standard low-level facilities of the UNIX operation system like sockets and the remote procedure call (RPC) are an efficient way to couple software systems. This approach has also proven to be very efficient in the GIS context for the direct communications between special applications (ALMS et al. 1998). However, this solution requires highly qualified programmers to absorb errors in the network communication. Furthermore, the modification and extension of the protocol requires costly changes in many pieces of the source code. To avoid these hardships for every pair of applications, we need a kind of standard distributed computing platform.

7.3.3 CORBA Communication Platform

The Common Object Request Broker Architecture (CORBA) of the Object Management Group (OMG) offers such a standardized platform. The advantage of CORBA -presented as version 2.0 in 1995 and published as a revised edition in 1997 (OMG 1997a)- is that it delegates much of the tedious and error-prone complexity associated with the low-level socket programming to its reusable infrastructure. The client and server applications can be written in different high-level programming languages. A client usually sees the server objects as usual objects of the client programming language and the communication is seen as an usual operation invocation. Developers can focus on more essential issues than analysing diverse networking troubles as in the case of sockets. A disadvantage of CORBA is that high-level protocols like the "Internet inter-ORB protocol" (IIOP) of the OMG have to be used instead of directly and efficiently exchanging the data and methods in binary format.

CORBA essentially consists of three parts: the Object Request Broker (ORB), a set of interfaces for the invocation of object-oriented operations and a set of object adaptors. The object interfaces have to be written in a common language to use the interfaces and adaptors. Furthermore, the languages used have to provide bindings for the common language. Examples for bindings to OMGs IDL are C, C++, Smalltalk and Java bindings.

The interfaces of the server objects are described in the IDL. These specifications are compiled with the IDL compiler at the "stub" on the client side and at the "skeleton surrogate" on the server side. The interfaces hide on both sides the data handling over the network as well as the interaction with the Object Request Broker (ORB) for the application programmer. The ORB is responsible for the creation of the communication on the server side. This includes the localization of a suitable target object and its activation. If neccesary, the sending of a request and the answer from the invocated side has to be handled.

The automatic generation of the stub and the skeleton surrogate ensures the correct interoperation between the client and the server components on different platforms. It excludes potential inconsistences between them and provides a higher degree of type safety than for bitstream-oriented sockets. The inevitable loss of efficiency in such universal middleware systems is compensated by the robustness and the global efficiency of the distributed system. A special problem, however, is to guarantee the persistency of CORBA objects (CHAUDHRI 1997).

7.3.4 Persistency of CORBA Objects

The CORBA-Standard (OMG 1997a) defines a flexible distribution model for transient objects. The situation with persistent objects, however, is more difficult. CORBA does not provide the support for persistent objects, but it offers an interface to integrate database management systems. The CORBA 2.2 specification (OMG 1998) allows to realize the integration of DBMSs in two ways: the first is called *Persistent Object Service* and the second implements an *Object Database Adapter*.

The Persistent Object Service (POS) is one of the standard object services specified in the OMG document (OMG 1997a). The objective of POS is to provide common interfaces for the mechanisms to manage persistent states of objects. Each object can choose by itself, if a client of an object should have knowledge about its persistent state. According to the knowledge of the author, hitherto no complete implementation of the POS standard exists.

Conversely, an Object Database Adapter (ODA) is an object adapter for objects stored in a database (OMG 1997a). It represents an extension of the usual Basic Object Adapter (BOA) by extra functions for management of persistent objects and provides a tight integration between the ORB and persistent object implementations. The same server can host regular CORBA transient objects, managed by the BOA, and persistent objects, managed by the BOA in conjunction with the ODA.

The BOA mediates between the ORB, the generated skeletons and the object implementation. It adapts the object implementation to the ORB specific interface with the help of skeletons or a dynamic skeleton interface. The target object implementations of ODA are persistently stored. A database adapter supplements the BOA by using loaders and filters. ODA is directly involved in the object reference generation, the object activation and re-gistration and the transaction management.

Since database management systems require that the access to persistent objects always must be performed within the transaction, ODA provides functions for management of transactions. ODA starts a transaction, if necessary, before invoking a method on a persistent object, and ends the transaction, implicitly or when instructed to do so, upon the completion of this call or some following call. The alternative would be to require an explicit management of transactions by the client application through the server IDL interface (REVERBEL 1996) or by an additional service (the Object Transaction Service).

ODA does not replace BOA completely, but it extends its properties. The same server can provide regular transient CORBA objects of the BOA as well as persistent objects that are managed by the BOA in connection with the Object Database Adapter.

To work around the weaknesses of existing Object Request Brokers (ORB) like the missing support of loaders and filters and to be more flexible, one's own implementations can be written analogously to the ODA. Then the CORBA objects are made persistently 'by hand'. In the following we will call the manual and the the automatic ORB techniques together the *wrapper approach.*

The critical points of the CORBA/database integration techniques can be summarized as follows:

1. *Object reference representation:* CORBA objects must be stored correctly in a persistent storage. In general the ODA approach allocates an object reference automatically. Conversely, the wrapper approach requires a programming interferer.
2. *Object activation:* In CORBA object activation and instantiation are synonymous. For persistent objects, however, these are two different aspects: persistent objects are instantiated once and can be activated or deactivated from that time on. The ODA approach simply loads a required persistent object into the server storage without generating any additional CORBA object. The wrapper approach is more complicated and requires a programming interferer for the object activation.
3. *Object deactivation:* The ODA approach does not generate a CORBA object. Thus no persistent object constructor is invocated. The server storage is managed by the database management system. The wrapper approach again requires a programming interferer with explicit storage management.
4. *Transaction management:* In the ODA approach transactions are successfully committed or aborted with an operation or phase approach. The wrapper approach is more flexible, because the programmer has a number of options to implement the transaction mechansims.
5. *Client interface:* The client should have the possibility to manage persistent objects. In the ODA and the wrapper approach the client usually does not have such mechanisms. Although in POS the standard client interface has been specified, it is not qualified for practical use.

The wrapper technique has proven to be more complex, but altogether it provides a better performance. The unsatisfactory ODA performance is caused by hidden costs, such as the allocation of a persistent CORBA object to an auxiliary CORBA object during the loading of a persistent object into the server's memory. However, the ODA approach better meets the CORBA specification. If the integration of existing applications with CORBA is required, it can be suitable to implement one's own approach. The ODA approach can be well used, if a closer integration between the ORB and the OODBMS is required. Database properties like the concurrency control can be used to achieve a corresponding object persistence. The wrapper approach is preferred for distributed environments with heterogeneous CORBA clients.

7.4 Implementation of a Component-Based Geoinformation System

In the following example the software components consist of application-specific classes of the geological and geophysical application (client components) and of GeoToolKit classes (database server component). Each component provides its specific object interface. The geological component (GOCAD), the 3D modelling tool developed at Nancy, France (MALLET 1992a), has been installed in Bonn[1] and the geophysical component in Berlin (IGMAS)[2]. The first two components have been used as an already existing application. The GeoToolKit component, however, has been newly developed[3]. The granularity of the selected components has been determined by the requirements of different geoscientific disciplines and by the availability of existing high-performance software. GeoToolKit served as a library for persistent geometric 3D datatypes and spatial access methods like the R*-Tree (BECKMANN et al. 1990; WAGNER 1998). The client components can store objects in GeoToolKit and access them again.

7.4.1 Access to an Open Geodatabase

One of the most important goals during the conception of a distributed geoinformation system is the development of an open geodatabase. The access has to be provided from different clients at heterogeneous hardware platforms and systems environments at the same time. Applications have to access the database with an *Application Programming Interface (API)*.

We show with the application of chapter 6.5 how geological and geophysical 3D modelling tools can access an open database for spatial data (Fig. 7.2) and how they can store data consistently in the database, respectively.

The open data exchange with a common database enables the user to deal with interdisciplinary evaluation of geological and geophysical model parameters under a common 3D model. This leads to a better understanding of geological and geophysical facts in the respective examination area.

Figure 7.2 shows the component diagram for a distributed geological-geophysical application between the geological 3D modelling tool GOCAD (MALLET 1992a), the geophysical IGMAS tool (GÖTZE and LAHMEYER 1988) and GeoToolKit (BALOVNEV et al. 1997a, 1998a) introduced in chapter 5.2. In the programme system IGMAS the geoscientific interpretation takes place. Data gained from wells and sections are compared with geological model conceptions.

[1.] Group of Agemar Siehl, Geological Institute, University of Bonn, Germany.
[2.] Group of HaJo Götze, Geophysical Institute, FU Berlin, Germany.
[3.] Group of Armin B. Cremers, Institute of Computer Science III, University of Bonn, Germany.

Chapter 7. Systems Integration

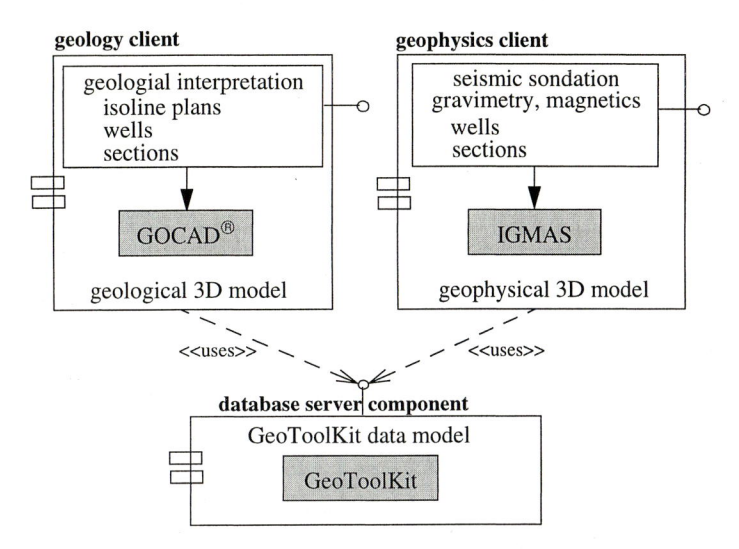

Fig. 7.2 Component diagram for a distributed geological-geophysical application on top of commonly managed geodata

In GOCAD the geological 3D model is interactively designed beginning with the geological interpretation of the wells and the construction of the sections and the stratum and fault surfaces. The data exchange with the data management component in GeoToolKit provides shared geo-objects.

7.4.2 CORBA-Based Solution

We introduce a CORBA-based solution that couples GeoToolKit with the geological and the geophysical 3D modelling tool. To make available GeoToolKit-based databases from the CORBA environment, the wrapper technique is used (ROTH and SCHWARZ 1997). The remote clients can interact with the wrapper by the IDL interface. The wrapper communicates with the GeoToolKit database and takes over the role of a remote representer for the clients at the server side.

The corresponding GeoToolKit wrapper uses CORBA as transparent network access between the different operation systems[1]. The wrapper enables the access to the remote CORBA clients GOCAD and IGMAS. Corresponding wrappers on the client side hide implementation details and supply the developer with an ODMG with a compatible API for the transparent access to the database (Fig. 7.3).

---

[1]. In the application example the platforms used are SUN and SGI workstations.

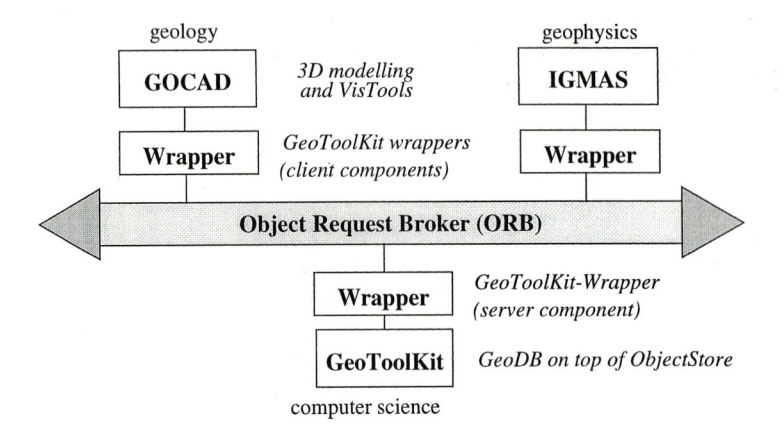

Fig. 7.3 Schema of the component-based GIS architecture

The wrapper around the GeoToolKit consists of a library of C++ classes in the upper layer of the OODBMS. The GeoToolKit class hierarchy in CORBAs IDL has been reproduced, trying to keep the interfaces as close as possible to those of the original C++ (Fig. 7.4). However, the IDL interfaces turned out to be significantly shorter, because local efficiency issues that are essential for the higher performance computations in the C++ environment were no longer relevant on the server side: the purchase of the optimization is insignificant in comparison with the total expenses for remote procedure calls. For example, the original GeoToolKit interface offers a bucket of functions implementing the same operation which are responsible for the efficient static polymorphic binding. The wrapper can do this internally without appreciable losses of efficiency. Figure 7.4 shows examples for the classes of the IDL interface for the GeoToolKit.

Figure 7.4 shows the realization of the container classes *GTK_Space* and *GTK_SpatialObject* inclusive of spatial predicates and operations. Furthermore, the class *GTK_Line* which is inherited by the *GTK_SpatialObject* is specified as an IDL interface.

```
exception GTK_ObjectExists {};              interface GTK_SpatialObject : GTK_Object {
exception GTK_ObjectNotFound {};
   ...                                          // spatial predicates
interface GTK_Space : GTK_Object {            boolean intersect (in GTK_SpatialObject
                                                 ...
// operations on spatial objects              // spatial operators

void insert (in GTK_SpatialObject             GTK_BoundBox getBoundBox ();
   raises(reject);                            GTK_SpatialObject
                                                 intersection(in GTK_SpatialObject
void remove (in GTK_SpatialObject
   raises (GTK_ObjectNotFound);               };

GTK_SpatialObject                             interface GTK_Line : GTK_SpatialObject {
   retrieve (in GTK_BoundBox bb);
                                              // functions specific for lines
GTK_Space                                        ...
   intersect (in GTK_SpatialObject            };
   ...
};
```

Fig. 7.4 IDL interface for the GeoToolKit (slightly changed from: BALOVNEV et al. 1998b)

To ensure the persistency of GeoToolKit objects, we choose a selection of the three approaches: Object Database Adapter (ODA), Wrapper and Persistent Object Service (POS) mentioned above. ODA is completely responsible for the management of persistent objects within ObjectStore, which on the initial stages considerably reduced development expenses. The use of ODA also improves the portability of the GeoToolKit wrapper code. From POS the ideas of a client IDL interface can be taken over to enable the management of persistent objects. All the original GeoToolKit classes are encapsulated by corresponding CORBA compatible mediator classes. The same is true for their objects that possess a pointer to the object of the corresponding GeoToolKit class. To bind the mediator classes to the corresponding IDL generated classes, the TIE technique is used. Figure 7.5 shows a complete schema of references between the objects for the GeoToolKit *gtplane* class.

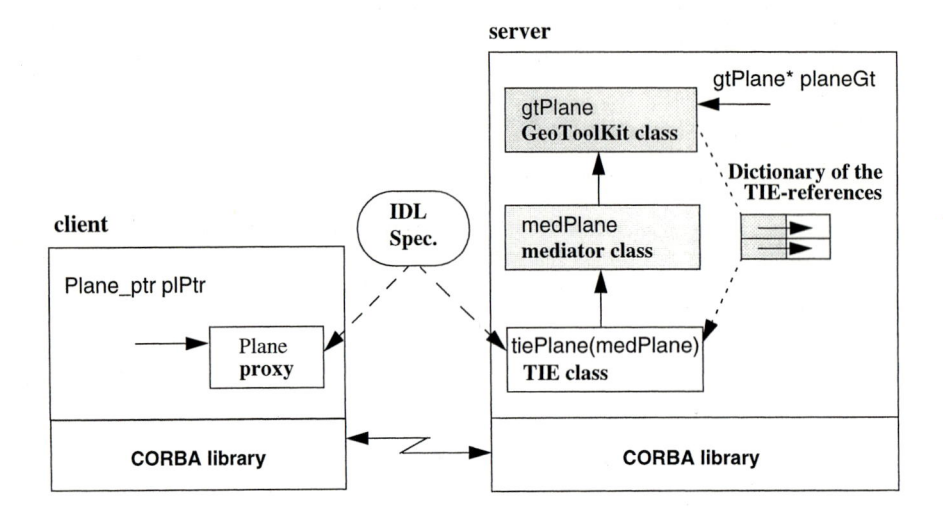

Fig. 7.5 General schema of the CORBA binding for the GeoToolKit classes, shown with the example of the *gtPlane* class

The mediator classes are responsible for the object generation and for the correct point initialization. Each mediator class has static functions that generate a triple of objects: the mediator object, and the corresponding TIE and GeoToolKit objects, respectively. As it is not possible for the client in CORBA to invoke the static methods on the server side, a special transient "object factory" has to be introduced. In its IDL interface this object has special functions that allow the client to control the corresponding objects for each GeoToolKit type by its life cycles. In the static generation functions all communication objects get correct pointers. The TIE object stores a pointer to the mediator object and the mediator object handles a pointer to the corresponding GeoToolKit object.

In an OODBMS each access to a persistent object has to be executed within a transaction. Therefore the wrapper has a local transaction mechanism that can be implemented as an extension of the CORBA/OODBMS adapter (BALOVNEV et al. 1997b). A wrapper should provide for clients flexible modi for the transaction control including multiple operation invocations.

A client communicates with server objects by CORBA object references. Thus a CORBA object reference identifies an object in the distributed system. If an object reference points to the address space of a client, CORBA generates a so called proxy object that stands for the remote mediator object. CORBA delegates the function calls on the proxy object to the corresponding functions in the mediator object. The most simple way to get a reference for an object is to use the binding mechanism of CORBA. Figure 7.6 shows an example for the client code. In the example a spatial retrieval is executed with a specified bounding box for geological objects (stratigraphic units). The corresponding GeoToolKit operation can be invoked by GOCAD and IGMAS, respectively.

Chapter 7. Systems Integration 177

```
#include <GTK_h.hh>
    ...
try {
    // get the reference to the factory and open database
    GTK_Factory_var gtk - GTK_Factory::_bind (":GTK", host);
    GTK_Database_var db = gtk->openDatabase ("Lower-Rhine-Basin");

    GTK_Point_var p1 = GTK_Point::_narrow (gtk->createObject(db, "Point"));
    ... // create three points specifiying a vertical plane
    GTK_Plane_var p1 = GTK_Plane::_narrow (gtk->createObject(db, "Plane"));
      p1->set_Point (p1, p2, p3);

    GTK_BoundBox_var bb = GTK_BoundBox::_narrow(gtk->createObject (db, "BoundBox"));
    bb->set_Double (250000, 560000, -1000, 254000, 568000, 100);

    GTK_Space_var layers = GTK_Space::_narrow (gtk->findObject (db, "StratigraphicUnits"));
    GTK_Space var rs = layers->retrieve (bb);
    GTK_SpatialObject_var so = rs->intersection (p1);

    ...
}; catch (...) {
    ...
};
```

Fig. 7.6 Example of the code for a client: retrieval of spatial objects in *spaces* (from: BALOVNEV et al. 1998b)

The parameters of the function for the binding mechanisms usually contain information about the location of the required mediator object in the distributed system. The parameters are composed by the name of the object, the server at which the object runs and the name of the host for the server. The result value then is a reference to the proxy object in the address space of the client. For access to an attribute or to an operation which is associated with an object, the client should invoke the corresponding member function of the proxy object that is returned by the binding function. The proxy addresses this C++ invocation over the network back to the corresponding member function of the mediator object.

Persistent objects of OODBMSs like ObjectStore usually are identified by their named root objects. The wrapper provides the possibility to generate a name for each GeoToolKit object. However, usually spatial objects are not maintained as separate database entities with external names. The triangles of a discretized surface of a geological surface collection, for example, should not be identified by their names. An object without root can only be found in the database with other objects that contain a pointer to this object. That is why spatial objects typically are encapusulated in large collections that are known as *spaces* in GeoToolKit. The number of diverse spaces in a database is usually not large. The costs are low to access their names. As a rule spaces have unique names and therefore they serve typically as entry points in spatial databases. The number of spatial objects is usually extremely large. Typically a separate spatial object is accessed indirectly via diverse retrieval functions associated with the space.

The coupling of the geological and the geophysical component via an open geodatabase enables the exchange of objects and the additional processing of objects that are generated or pre-processed in the other component, respectively.

The advantages of the systems integration are obvious. Applying the proposed approach, both geology and geophysics gain benefit, which is two-sided. The result is a data and methods integration. Furthermore, the object-oriented modelling approach has been proven to be a suitable instrument for the spatial representation of different geodata.

The coupled approach of geological and geophysical 3D modelling tools has lead to a better understanding of sub-surface geology. Geology has mainly gained by the coupling with the geophysical 3D modelling tool, because the interactively modelled bodies can now be checked in IGMAS according to their densities or their position and geometry, respectively. The 3D geometry has to be corrected in GOCAD, if the computed gravity field strongly varies from the expected values. If necessary, the density values for stratigraphical and facies bodies also have to be re-computed. This directly leads to a qualitative improvement of the geologically defined 3D model. Conversely in the 3D modelling tool IGMAS only a rudimentary interactive processing of 3D geometries is provided. That is why geophysics profits from the coupling to GOCAD particularly by the additional GOCAD methods for a fine tuning of the geometry and topology. At the same time corrections of the density model can be done. Thus we have succeeded in bringing together geological and geophysical modelling. The coupling with an open, object-oriented database resulted in a synergy effect for the geological-geophysical processing of digital data. Furthermore, the computer scientists could present an ambitious application for component-based software systems.

Chapter 8

Outlook

Obviously there are still open questions that can only be answered by further practical experience. Examples of such problems are:

1. Should the component-based approach of systems architecture also be used to let the geoscientist "tailor" her/his own GIS with freely configurable components? Is this realistic?
2. How can the spatial and temporal techniques of database access be used for the support of picture and raster-based multi media applications? How can the object-based access be realized to achieve the access to a specified target in an animation or in a video sequence?

The first point has two facets: on the one hand, software engineering has made progress with component-based system architectures. It is ready to implement flexible and freely configurable systems. On the other hand, the expectations of few geoscientists are still very high towards new system developments. Sometimes detailed functionality of the user-interface seems to be more important than a flexible and extensible systems approach. Anyway, the future needs an ongoing process of close cooperation between geoscientists and computer scientists. Otherwise the progress achieved in the last years will soon be forgotten and new cooperating groups will have to start right from the beginning.

The second point is of general interest, not only for the geosciences but also for video-on--demand applications in different fields and for citizen-based information systems. Efficient object-based spatial and temporal access methods will be of great importance, because very large raster sequences have to be transferred efficiently via internet to support multi media databases. A new field of research could be created by combining object recognition strategies with object-based multi-dimensional acceess methods.

References

ABEL DJ (1989) SIRO-DBMS: A Database Tool-kit for Geographical Information Systems. In: Intern. Journal of Geographical Information Systems 3, pp 291-301

ABEL D, KILBY PJ, DAVIS JR (1994) The Systems Integration Problem. In: Intern. Journal of Geographical Information Systems, vol. 3, pp 291-301

ABEL D, Ooi BC (Eds., 1993) Advances in Spatial Databases. In: proceedings of the Third Intern. Symposium SSD'93, LNCS, Springer Verlag, Berlin et al.

ABEL DJ, WILSON, MA (1990) A Systems Approach to Integration of Raster and Vector Data and Operations. In: proceedings of the 4th Intern. Symposium on Spatial Data Handling, Zürich, Vol. 2, pp 559-566

ABEL DJ, YAP S, WALKER G, CAMERON MA, ACKLAND RG (1992) Support in Spatial Information Systems for Unstructured Problem-Solving. In: proceedings of the 5th International Symposium on Spatial Data Handling (SDHS), Charleston, South Carolina, pp 434-443

ACM (1990) Special Issue on Heterogeneous Databases, ACM Computing Surveys 22 (3)

AdV (1989) ATKIS documentation, Arbeitsgemeinschaft der Vermessungsverwaltungen der Länder der Bundesrepublik Deutschland (AdV)

ALLEN JF (1983) Maintaining Knowledge about Temporal Intervals. Communications of the ACM. 26(11), Nov., pp 832-843

ALLEN JF (1984) Towards a General Theory of Action and Time. In: Artificial Intelligence, 23(2), pp 123-154

ALMS R, BALOVNEV O, BREUNIG M, CREMERS AB, JENTZSCH T, SIEHL A (1998) Space-Time Modelling of the Lower Rhine Basin supported by an Object-orien-ted Database. In: Physics and Chemistry of the Earth, Vol.23, No.3, Elsevier Science Ltd., pp 251-260

ALMS R, KLESPERS C, SIEHL A (1994) Geometrische Modellierung und Datenbankentwicklung für dreidimensionale Objekte. In: IfAG Nachrichten aus dem Karten- und Vermessungswesen, series 1

ANDERL R, SCHILLI B (1988) STEP - Eine Schnittstelle zum Austausch integrierter Modelle. In: Weber HR (ed.). CAD-Datenaustausch und -Datenverwaltung, Springer-Verlag, Berlin Heidelberg

ARCHONTAS D (1997) Datenbankunterstützung für die klimatische und ökologische Auswertung paläontologischer Daten in einem Bohrdatenverwaltungssystem. Diploma thesis, Oct. 1997, Institute of Computer Science III, University of Bonn, Germany, 54p.

ATKINSON M, BANCILHON F, DEWITT D, DITTRICH K, MAIER D, ZDONIK S (1989) The Object-Oriented Database Manifesto

BALOVNEV O, BREUNIG M (1997) Erste Versuche zur Modellierung der Zeit als Anwendung eines GeoToolKits. In: proceedings of workshop "Zeit als weitere Dimension in Geo-Informationssystemen", 29.-30. Sept., University of Rostock, Institute of Geodesy and Geoinformatics, techn. report no. 7, pp107-114

BALOVNEV O, BREUNIG M, CREMERS AB (1997a) From GeoStore to GeoToolKit: the second step. In: proceedings of the 5th Intern. Symposium on Spatial Databases, Berlin, LNCS No. 1262, Springer, Berlin et al., pp 223-237

BALOVNEV O, BREUNIG M., CREMERS AB, PANT M (1998a) Building Geoscientific Applications on Top of GeoToolKit: a Case Study of Data Integration. In: Proceedings of SSDM'98 Symposium on Scientific and Statistical Databases, Capri, IEEE, pp 260-268

BALOVNEV O, BERGMANN A, BREUNIG M, CREMERS AB, SHUMILOV S (1998b) A CORBA-based Approach to Data and Systems Integration for 3D Geoscientific Applications. In: proceedings of the 8th Intern. Symposium on Spatial Data Handling, Vancouver, Canada, pp 396-407

BALOVNEV O, BREUNIG M, CREMERS AB, SHUMILOV S (1997b) GeoToolKit: Opening the Access to Object-oriented Geodata Stores. In: proceedings of the Intern. Conference on Interoperating Geographic Information Systems, Santa Barbara, CA, USA, Dec. 1997, 10 p.

BANCILHON F, CLUET S, DELOBEL C (1989) A Query Language for the O_2 Object-oriented Database System. In: proceedings of the 2nd Workshop on Database Programming Languages, Salishan, Oregon, USA

BARTELME N (1995) Geoinformatik, Springer, Berlin et al., 411p.

BATORY D, BARNATT J, GARZA J, SMITH K, TSUKUDA K, TWICHELL C, WISE T (1986) GENESIS: A Reconfigurable Database Management System, tech. rep. 86-07, Dept. of Computer Science, University of Texas at Austin, USA

BAYER R, MC CREIGHT E (1972) Organization and Maintenance of Large Ordered Indexes. In: Acta Informatica 1, Springer Verlag, Berlin et al., pp 173-189

BECKER L, VOIGTMANN A, HINRICHS K (1996) Developing Applications with the Object-oriented GIS-Kernel GOODAC. In: proceedings of the 7th Intern. Symposium on Spatial Data Handling, Delft, 17p.

BECKMANN N, KRIEGEL H-P, SCHNEIDER R, SEEGER B (1990) The R*-tree: An Efficient and Robust Access Method for Points and Rectangles. In: proceedings ACM SIGMOD, Atlantic City, N.Y., pp 322-331

BILL R (1992) Multi-Media-GIS - Definition, Anforderungen und Anwendungsmöglichkeiten, ZfV, 7/1992, pp 407-416

BILL R (1996) Grundlagen der Geo-Informationssysteme, Vol. 2. Wichmann Verlag, Karlsruhe, 463 p.

BILL R (1997) Ed., proceedings of the workshop "Zeit als weitere Dimension in Geo-Informationssystemen", 29./30 Sept., Universität Rostock, 143p.

BILL R, FRITSCH D (1991) Grundlagen der Geo-Informationssysteme, Vol. 1. Wichmann Verlag Karlsruhe, 414p.

BISHR Y (1997) Semantic Aspects of interoperable GIS. In: ITC Publication Series, No. 56, Den Haag, 154 p.

BODE TH, BREUNIG M, CREMERS AB (1994) First Experiences with GEOSTORE, an Information System for Geologically Defined Geometries. In: Nievergelt J, Roos Th, Schek H-J, Widmayer P (Eds.), IGIS'94: Geographic Information Systems, proceedings of the Intern. Workshop on Advanced Research in Geograph. Inform. Systems, Monte Verita, Ascona, Schweiz, Feb. 28-March 4, LNCS 884, Springer, Berlin et al., pp 35-44

BODE TH, CREMERS AB, FREITAG J (1992) OMS - ein erweiterbares Objektmanagement-System. In: Objektbanken für Experten, GI Fachberichte, Springer Verlag, Berlin et al., pp 29-54

BOUILLE F (1976) Graph Theory and Digitization of Geological Maps. In: Computers & Geosciences, Elsevier Science, Oxford, Vol.2, pp 375-393

BOURSIER P, MAINGUENAUD M (1992) Spatial Query Languages: Extended SQL versus Visual Languages versus Hypermaps. In: Proceedings of the 5th Intern. Symposium on Spatial Data Handling, Charleston SC, IGU Commission of GIS, pp 249-259

BRANKI T, DEFUDE B (1998) Data and Metadata: Two-Dimensional Integration of Heterogeneous Spatial Databases. In: proceedings of the 8th Intern. Symposium on Spatial Data Handling, Vancouver, Canada, pp 172-179

BRASSEL K (1993) Grundkonzepte und technische Aspekte von Geographischen Informationssystemen. IJK, pp 31-52

BREUNIG M (1989) Der R^c-Baum: Implementierung und Evaluierung eines räumlichen Zugriffspfades. Diploma thesis, Department of Computer Science, Technical University of Darmstadt, Germany, 99p.

BREUNIG M (1996) Integration of Spatial Information for Geo-Information Systems, LNES No. 61, Springer Verlag, 171 p.

BREUNIG M (1998) GeoToolKit: Idea, Design and Implementation Aspects. In: Proceedings of the GOCAD-ENSG Conference on 3D Modelling of Natural Objects, Vol. 2, Nancy, 4-5 June, France, 10p.

BREUNIG M (1999) An Approach to the Spatial Data and Systems Integration for a 3D Geo-Information System. In: Special Issue on Systems Integration within the Geosciences. Computers & Geosciences, Vol. 25, No. 1, Elsevier Science, Oxford, pp 39-48

BREUNIG M, BODE TH, CREMERS AB (1994) Implementation of Elementary Geometric Database Operations for a 3D-GIS. In: Waugh Th, Healey R (Eds.), proceedings of the 6th Intern. Symposium on Spatial Data Handling, 5th - 9th September 1994, Edinburgh, Scotland, UK, pp 604-617

BREUNIG M, CREMERS AB, GÖTZE H-J, SCHMIDT S, SEIDEMANN R, SHUMILOV S, SIEHL A (1999) First Steps Towards an Interoperable GIS - An Example From Southern Lower Saxony. Physics and Chemistry of the Earth, Elsevier Science, Oxford, pp 179-189

BREUNIG M, KLETT M (1999) Object-oriented Modelling and Management of Geological and Geophysical Well Data in the Lower Rhine Embayment. Submitted for: Geolog. Rundschau, 15 p.

BREUNIG M, PERKHOFF A (1992) Data and System Integration for Geoscientific Data. In: 5th Intern. Symposium on Spatial Data Handling, Charleston, South Carolina, 3.-7. Aug. , pp 272-281

BRINKHOFF TH, HORN H, KRIEGEL H-P, SCHNEIDER R (1993) Eine Speicher- und Zugriffarchitektur zur effizienten Anfragebearbeitung in Geo-Datenbanksystemen. In: proceedings of BTW'93

BRUZZONE E, DE FLORIANI L, PELLEGRINELLI (1993) A Hierarchical Spatial Index for Cell Complexes. In: proceedings of SSD'93 Singapore, LNCS 692, Springer, Berlin et al., pp 105-122

BUEHLER K, MC KEE, L (eds.)(1996) The OpenGIS Guide - Introduction to Interoperable Geoprocessing, Open Geodata Interoperability Specification (OGIS). Open GIS Consortium, Inc., http://www.ogis.org/stuff.html

BURNS K L (1975) Analysis of Geological Events. In: Mathematical Geology, Vol. 7, No. 4, 1975

CAREY MJ, DEWITT DJ, FRANK D, GRAEFE, G, MURALIKRISHNA M, RICHARDSON JE, SHEKTIA EJ (1988) The Architecture of the EXODUS Extensible DBMS. In: Stonebraker M (Ed.), Readings in Database Systems, Morgan Kaufman

CERI S, GOTTLOB G (1985) Translating SQL Into Relational Algebra: Optimization, Semantics, and Equivalence of SQL Queries. In: IEEE Transactions of Software Engineering SE-11, pp 324-345

CHAUDHRI AB (1997): Workshop Report on Experiences Using Object Management in the Real World. OOPSLA'97 workshop "Experiences Using Object Data Management", Atlanta Ga, USA, Oct. 1997

CLEMENTINI E, DI FELICE P (1994) A Comparison of Methods for Representing Topological Relationships. In: Information Sciences 80, pp 1-30

CODD EF (1970) A Relational Model of Data for Large Shared Data Banks. In: Communications of the ACM, 13, pp 377-387

CONREAUX S, LÉVY B, MALLET J-L (1998) A Cellular Topological Model Based on Generalized Maps. In: 3D Modelling of Natural Objects - A Challange for 2000s, GOCAD/ENSG 3D Modelling Conference, Nancy, Vol.1, 17p.

COWEN DJ (1988) GIS versus CAD versus DBMS: What Are the Differences. In: Photogramm. Eng. Remote Sensing, 54, 11, Falls Church, VA, pp 1551-1555

CREMERS AB, KNIESEL G, LEMKE T, PLÜMER L (1992) Intelligent Databases and Interoperability. In: Belli F, Radermacher FJ (eds.), Industrial and Engineering Applications of Artificial Intelligence and Expert Systems, LNAI No. 604, Springer, Berlin et al.

CREMERS AB, BALOVNEV O, REDDIG W (1994) Views in Object-oriented Databases. In: proceedings of the Intern. Workshop on Advances in Databases and Information Systems ADBIS'94, Moscow ACM SIGMOD Chapter, May 23-26, Moscow

DANGERMOND J (1983) A Classification of Software Components Commonly Used in Geographical Information Systems. In: Marble D, Calkins H, Peuquet D (Eds.), Basic Readings in Geographic Information Systems. SPAD Systems, Amherst, N.Y.

DANN R, SCHULTE-ONTROPP R (1989)   DIGMAP :   Das Software - System im Markscheidewesen der Ruhrkohle AG, Essen, Germany

DAVID B, RAYAL L, SCHORTER G, MANSART V (1993) GeO_2: Why Objects in a Geographical DBMS? In: proceedings of the 3rd Intern. Symposium on Large Spatial Databases, Singapore, 23rd - 25th June, pp 264-276

DAYAL U, MANOLA F, BUCHMANN A, CHAKAVARTHY D, GOLDHIRSCH D, HEILER S, ORENSTEIN J, ROSENTHAL A (1987) Simplifying Complex Objects: The PROBE Approach to Modelling and Querying them. In: proceedings BTW'87

DE FLORIANI L, GATTORNA G, MARZANO P, PUPPO E (1994) Spatial Queries on a Hierarchical Terrain Model. In: proceedings of the 6th Intern. Symposium on Spatial Data Handling, Edinburgh, pp 819-834

DE HOOP S, VAN DER MEIJ L, VAN HEKKEN M, VIJLBRIEF T (1994) Integrated 3D Modelling within a GIS. In: proceedings of the Intern. GIS Workshop "Advanced Geographic Data Modelling", AGDM'94, Delft

DEUX O (1990) The story of O_2. In: IEEE Transactions on Knowledge and Data Engineering, 2(1), pp 91-108

DEVOGELE T, PARENT C, SPACCAPIETRA S (1998) On spatial database integration. In: Intern. Journal of Geographic Information Systems, vol. 12, no.4

DIKAU R (1992) Aspects of Constructing a Digital Geomorphological Base Map. In: From Geoscientific Map Series to Geo-Information Systems. In: Geologisches Jahrbuch A122, Hannover, pp 357-370

DITTRICH KR (1986) Object-Oriented Database Systems: The Notion and the Issue. In: Dittrich KR, Dayal U (Eds.), Intern. Workshop on Object-Oriented Database Systems, September 23-26, Pacific Grove, California, USA, 2-4, proceedings IEEE-CS 1986

DOERPINGHAUS F (1989) Implementation eines räumlichen Zugriffspfades nach der Grid-File Methode. Study thesis, Department of Computer Science, Technical University of Darmstadt, Germany, 92p.

EBBINGHAUS J, HESS G, LAMBACHER J, RIEKERT W-F, TROTZKI T, WIEST G (1994) GODOT: Ein objektorientiertes Geoinformationssystem. In: Hilty LM, Jaeschke A, Page B, Schwabl A (Eds.), proceedings of 8th Symposium für den Umweltschutz, Hamburg, Metropolis-Verlag, Marburg, pp 351-360

EDELSBRUNNER H (1992) Weighted Alpha Shapes. University of Illinois at Urbana-Champaign, Dep. of Computer Science, Urbana, Illinois, Report No. UIUCDCSR-92--1760 (UILU-ENG-92-1740), July, 15p.

EDELSBRUNNER H, KIRKPATRICK DG, SEIDEL R (1983) On the Shape of a Set of Points in the Plane: In: IEEE Transaction Infom. Theory, IT-29, pp 551-559

EDELSBRUNNER H, MÜCKE EP (1994) Three-dimensional Alpha Shapes. In: ACM Transactions on Graphics 13, pp 43-72

EGENHOFER MJ (1989) A Formal Definition of Binary Topological Relationships. In: Litwin W, Schek H-J (Eds.), Foundations of Data Organisation and Algorithms, proceedings FODO 1989, Paris LNCS 367, Springer, Berlin et al., pp 457-472

EGENHOFER MJ (1991) Reasoning About Binary Topological Relations. In: Advances in Spatial Databases, 2nd Symposium SSD'91, Zürich, LNCS No. 525, Springer, Berlin et al., pp 143-160

EGENHOFER MJ (1994) Spatial SQL: A Query and Presentation Language. In: IEEE Transactions on Knowledge and Data Engineering, Vol. 6, No. 1, Feb.

EGENHOFER MJ, AL-TAHA KK (1992) Reasoning about Gradual Changes of Topological Relationships. In: Frank AU, Campari I, Formentini U (Eds.), Theories and Models of Spatio-Temporal Reasoning in Geographic Space, LNCS 639, Springer, pp 196-219

EGENHOFER MJ, FRANK A (1988) Designing Object-Oriented Query Languages for GIS: Human Interface Aspects. In: proceedings of the 3rd Intern. Symposium on Spatial Data Handling, Sidney, pp 79-96

EGENHOFER MJ, FRANZOSA RD (1991) Point-set Topological Spatial Relations. In: Intern. Journal of Geographical Information Systems, 5(2), pp 161-174

EGENHOFER MJ, FRANZOSA RD (1995) On the Equivalence of Topological Relations. In: Intern. Journal of Geographical Information Systems, 9(2), pp 133-152

ERWIG M, SCHNEIDER M, GÜTING RH (1997a) Temporal and Spatio-temporal Data Models and their Expressive Power. In: technical report no. 225, Dec. 1997, Praktische Informatik IV, Fernuniversität Hagen, Germany, 21 p.

ERWIG M, GÜTING RH, SCHNEIDER M, VAZIRGIANNIS M (1997b) Spatio-temporal Data Types: an Approach to Modelling and Querying Moving Objects in Databases. Technical report No. 224, 1997, Pract. Comp. Sc. IV, Fernuniversität Hagen, Germany, 13 p.

FELLNER DW, FISCHER M, WEBER J (1993) CGI-3D - A 3D Graphics Interface, IAI-TR-95-x, Aug., Institute of Computer Science III, University of Bonn, Germany

FINKEL RA, BENTLEY JL (1974) Quad trees: A Data Structure for Retrieval on Composite Keys. In: Acta Informatica, Vol. 4, No. 1, pp 1-9

FISHER PF (1998) Geographic Information Science and Systems: A Case of the Wrong Me-taphor. In: proceedings of the 8th Intern. Symposium on Spatial Data Handling, Vancouver, Canada, pp 321-330

FITZKE J, RINNER C, SCHMIDT D (1997) GIS-Anwendungen im Internet. In: Geo-Informationssysteme 10, vol. 6, pp 25-31

FONG EN, GOLDFINE AH (1989) Special Report: Information Management Directions: The Integration Challenge. In: SIGMOD RECORD Vol. 18, No. 4, Dez, pp 41-43

FRANK AU (1982) MAPQUERY - Database Query Language for Retrieval of Geometric Data and its Graphical Representation. In: ACM Computer Graphics, Vol.3, No. 16, pp 199-207

FRANK AU (1994) Qualitative Temporal Reasoning in GIS-Ordered Time Scales. In: proceedings of the 6th Intern. Symposium on Spatial Data Handling, Edinburgh, pp 410-430

GAEDE V, GÜNTHER O (1998) Multidimensional Access Methods. In: ACM Computing Surveys, 30(2), June, pp 123-169

GAEDE V, RIEKERT W-F (1994) Spatial Access Methods and Query Processing in the Object-oriented GIS GODOT. In: proceedings of the Intern. GIS Workshop on Advanced Geographic Data Modelling (AGDM-94), Delft, 12. - 14. Sept.

GERRITSEN BHM (1998) On the use of α-Complexes in Subsurface Modelling. In: 3D Modelling of Natural Objects - A Challange for 2000s, GOCAD/ENSG 3D Modelling Conference, Nancy, 27p.

GOCAD (1996) GOCAD Users' Manual. - Draft Copy for Release 1.3.3, 924 S., Nancy (ENSG)

GOCAD (1999) http://www.ensg.u-nancy.fr/GOCAD/gocad.html

GOODCHILD MF (1985) Geographic Information Systems in Undergraduate Geography: A Contemporary Dilemma. In: the Operational Geographer, Vol. 8, pp 34-38

GOODCHILD MF (1990) Spatial Information Science. Keynote Address. In: proceedings of the 4th. Intern. Symposium on Spatial Data Handling. Vol. 1, Zürich, pp 3-12

GÖPFERT W (1991) Raumbezogene Informationssysteme, 2. edition. Wichmann Verlag, Karlsruhe, 318p.

GÖTZE H-J (1978) Ein numerisches Verfahren zur Berechnung der gravimetrischen Feldgrößen drei-dimensionaler Modellkörper.- Arch. Met. Geoph. Biokl., Wien, series A, pp 195-215

GÖTZE H-J (1984) Über den Einsatz interaktiver Computergraphik im Rahmen 3-dimensionaler Interpretationstechniken in Gravimetrie und Magnetik. Habilitation thesis, TU Clausthal, Clausthal, 121p.

GÖTZE H-J, LAHMEYER B (1988) Application of Three-dimensional Interactive Modelling in Gravity and Magnetics. In: Geophysics, Vol. 53, No. 8, pp 1096-1108

GRAPE (1997) Graphics Programming Environment - Manual Version 5.3; Institut für angewandte Mathematik, SFB 256, University of Bonn, Germany, http://www.iam.uni-bonn.de/sfb256/grape

GRASS (1993) GRASS 4.1 Reference Manual, U.S. Army Corps of Engineers, Construction Engineering Research Laboratories, Champaign, Illinois.

GREVE K, HEISS M, WESELOH R (1997) Umweltinformationssysteme als Grundlage des Umweltschutzes. In: GIS, 1/97, pp 6-11

GRIEBEL M (1994) Domain-oriented Multilevel Methods. In: Keyes D, Xu J (Eds): Contemporary Mathematics 180, DDM7, American Mathematical Society, pp 223-229

GRUGELKE G (1986) Benutzerhandbuch THEMAK2, Version 2.0, Freie Unviersität Berlin, Germany

GRÜNBAUM (1967) Convex Polytopes. Wiley, New York, 1967

GRÜNREICH D (1992) ATKIS - A Topographic Information System as a Basis for a GIS and Digital Cartography in West Germany. In: Geolog. Jahrbuch, A122, pp 207-216

GUENTHER O, BUCHMANN A (1990) Research Issues in Spatial Databases. In: IEEE Bulletin on Data Engineering 13(4), pp 35-42

GÜNTHER O, LAMBERTS J (1992) Object-oriented Techniques for the Management of Geographic and Environmental Data, FAW Technical Report 92023, Sept.

GÜTING RH (1988) Geo-relational Algebra: A Model and Query Language for Geometric Database Systems. In: Schmidt JW, Ceri S, Missikoff M (Eds.), Advances in Database Technology - EDBT '88. Proceedings of the Intern. Conference on Extending Database Technology, Venice, Italy, March , pp 506-527

GÜTING RH (1989) Gral: An Extensible Relational Database System for Geometric Applications. In: proceedings of the 15th Intern. Conference on Very Large Data Bases, Amsterdam, Netherlands, August, pp 22 - 25

GÜTING RH (1992) Datenstrukturen und Algorithmen, Teubner, Stuttgart, Germany, 308p.

GÜTING RH (1994) An Introduction to Spatial Database Systems. In: the VLDB Journal,Vol.3, No. 4, Oct.

GÜTING RH, SCHNEIDER M (1993) Realm-Based Spatial Data Types: The ROSE Algebra, Computer Science Reports No. 141, 3/1993, Fernuniversität Hagen, Germany

GUTTMAN A (1984) R-Trees: A Dynamic Index Structure for Spatial Searching: in: proceedings of the Annual Meeting ACM SIGMOD, Boston (MA), pp 47-57

HACK R, SIDES E (1994) Three-dimensional GIS: Recent Developments, ITC Journal, 1994-1, Delft

HAAS LM, CODY WF (1991) Exploiting Extensible DBMS in Integrated Geographic Information Systems. In: Günther O, Schek H-J (Eds): Advances in Spatial Databases. Proceedings of the 2nd Symposium SSD, Zürich, Switzerland, Lecture Notes in Computer Science No. 525, Springer, Berlin et al., pp 423-450

HAZELTON N, LEAHY F, WILLIAMS I (1990) On the Design of Temporally-Referenced 3D-Geographical Information Systems: Development of Four-Dimensional GIS. In: proceedings of GIS/LIS'90. Volume 1, pp 357-372

HÄRDER T, MEYER-WEGENER K, MITSCHANG B, SIKELER A (1987) PRIMA - a DBMS Prototype Supporting Engineering Applications. In: proceedings of the 14th Intern. Conference on Very Large Data Bases, Aug. 29 - Sept. 1, Los Angeles

HEALEY G, WAUGH TH (1994) Advances in GIS Research. In: proceedings of the 6th Intern. Symposium on Spatial Data Handling, 5th-9th Sept, Edinburgh, 2 Vols.

HEARNSHAW HM, UNWIN DJ (eds., 1994) Visualization in Geographical Information Systems. John Wiley & Sons, Chichester et al.

HEINKE B (1997) Entwurf und Implementierung der Verwaltung von Volumina im GeoToolKit. Diploma thesis, Institute of Computer Science III, University of Bonn, Germany, 82p.

HENRICH A, SIX H-W, WIDMAYER P (1989) The LSD Tree: Spatial Access to Multidimensional Point and Non Point Objects. In: proceedings of VLDB, Amsterdam

HERRING JR (1987) TIGRIS: Topologically Integrated Geographic Information System. In: proceedings of Auto-Carto 8, pp 282-291

HEUER A (1992) Objektorientierte Datenbanken, Addison-Wesley, Bonn et al., 628p.

HILGER R (1998) Überprüfung räumlicher Integrität für ein 3D-GIS am Beispiel von GeoStore. Diploma thesis, Institute of Computer Science III, University of Bonn, Germany, 101p.

HIRTLE SC, FRANK AU (1997)(eds.) Spatial Information Theory - a Theoretical Basis for GIS, proceedings of COSIT'97, LNCS No. 1329, Springer, Heidelberg et al.

IEEE (1991) IEEE Computer. Special Issue on Heterogeneous Distributed Systems 24(12), Dec.

JAESCHKE G, SCHEK H-J (1981) Remarks on the Algebra of Non-First-Normal-Form Relations. In: proceedings of the 1st. ACM SIGACT/SIGMOD Symposium on Principles of Database Systems, Los Angeles, Ca.

JONES CB (1989) Data Structures for Three-dimensional Spatial Information. In: Geographical information systems, Vol. 3, No. 1, Jan-March, Taylor & Francis, pp 15-31

KAINZ W (1991) Spatial Relationships - Topology versus Order. In: proceeding of the 4th Intern. Symposium on Spatial Data Handling, Zürich, pp 814-819

KELK B (1992) 3-D Modelling with Geoscientific Information Systems: the Problem. In: A. K. Turner (ed.), Three-Dimensional Modelling with Geoscientific Information Systems, NATO ASI 354, Kluwer Academic Publishers, Dordrecht, pp 29-38

KLESPER C (1994) Die rechnergestützte Modellierung eines 3D-Flächenverbandes der Erftscholle (Niederrheinische Bucht). PhD thesis, Geological Institute, University of Bonn. In: Berliner Geowissenschaftliche Abhandlungen, series B, vol. 22, 51 fig., Berlin, 117 p.

KLETT M, SCHÄFER A (1996) Interpretation des Tertiärs der Niederrheinischen Bucht anahand von Bohrungen. - abstract. In: proceedings of 148. Jahrestagung der Deutschen Geologischen Gesellschaft, 1.-3.Oct, Bonn, Germany

KOCKEL F et al. (1996) Geotektonischer Atlas von Nordwest-Deutschland, BGR, Hannover, Germany

KOLLARITS S (1990) SPANS 5.0, Benutzeranleitung FMM - TYDAC, Salzburg, Austria

KOSCHEL A, KRAMER R, THEOBALD D, VON BÜLTZINGSLOEWEN G, HAGG W, WIESEL J, MÜLLER M (1996) Evaluierung und Einsatzbeispiele von CORBA-Implementierungen von CORBA-Implementierungen für Umwelinformationssysteme. In: proceedings of the 10th Symposium Informatik für den Umweltschutz, Hannover, Metropolis-Verlag, pp 190-200

KRAAK M-J, VERBREE F (1992) Tetrahedrons and Animated Maps in 2D and 3D Space. In: proceedings of the 5th Intern. Symposium on Spatial Data Handling. Charleston, South Carolina, 1992, pp 63-71

LEE YJ, LEE SJ, CHUNG CW (1997): Object Decomposition for Spatial Query Processing. In: Intern. Journal of Information Technology, Vol. 3, No. 1, World Scientific Pub. Com., 35-62

LIENHARDT P (1989) Subdivisions of Surfaces and Generalized Maps. Proc. Eurographics '89, Hamburg, pp 439-452

LIENHARDT P (1994) N-Dimensional Generalized Combinatorial Maps and Cellular Quasi-Manifolds. In: Intern. Journal of Computational Geometry and Applications, Vol. 4, no. 3, pp 275-324

LIENHARDT P, BECHMANN D, BERTRAND Y (1997) Four-Dimensional Modelling with Generalized Maps. In: proceedings of the GOCAD Workshop, Nancy, June 16th-20th

LINNEMANN V, KÜSPERT K, DADAM P, PISTOR P, ERBE R, KEMPER A, SÜDKAMP N, WALCH G, WALLRATH M (1988) Design and Implementation of an Extensible Database Management System Supporting User Defined Data Types and Functions. In: proceedings of the 14th Intern. Conference on Very Large Data Bases, Los Angeles

LOMET D, WIDOM J (1995) (Eds.) Special Issue on Materialized Views and Data Warehousing. IEEE Data Engineering Bulletin 18 (2), June

LYNX (1996) LYNX, Version 4.4, User Documentation, Volume I: User Guide, Lynx Geosystems Inc., Vancouver, Canada

MALLET JL (1992a) GOCAD: a Computer Aided Design Programme for Geological Applications. In: Turner AK (Ed.), Three-Dimensional Modelling with Geoscientific Information Systems. In: proceedings of NATO ASI 354, Kluwer Academic Publishers, Dordrecht, pp 123-142

MALLET JL (1992b) Discrete Smooth Interpolation in Geometric Modelling. In: Computer Aided Design, 24, 4, pp 178-192

MALLET JL (1998a) (ed.) 3D Modelling of Natural Objects - A Challenge for 2000s, GOCAD/ENSG 3D Modelling Conference, Nancy, 3 Vols.

MALLET JL (1998b) Modelling Natural Objects: the gOcad approach. In: 3D Modelling of Natural Objects - A Challenge for 2000s, GOCAD/ENSG 3D Modelling Conference, Nancy, Vol.1, 17p.

MÄNTYLA M (1988) An Introduction to Solid Modelling, Computer Science Press, 401p.

MC GAUGHEY J (1997) Geological Modelling at a Major Mining Company: Past, Present and Future. In: Abstracts of the European Research Conference on Space-Time Modelling of Bounded Natural Domains: Virtual Environments for the Geosciences, 9.-14. Dec., Kerkrade, Netherlands

MC CORMIC BH, DeFANTI TA., BROWN MD (1987)(eds.) Visualization in Scientific Computing. Special issue ACM SIGGRAPH, Computer Graphics 21 (6)

MEIER A (1986) Methoden der grafischen und geometrischen Datenverarbeitung.Teubner Stuttgart

MOREHOUSE S (1985) ARC/INFO a Geo-relational Model for Spatial Information. In: proceedings Auto Carto 7, Washington D.C., pp 338-357

MORTON GM (1966) A Computer Oriented Geodetic Database and a New Technique in Filesequencing, IBM, Ottawa, Canada

MOWBRAY T J, ZAHAVI R (1995) THE ESSENTIAL CORBA - Systems Integration Using Distributed Objects, Object Management Group, John Wiley & Sons, New York et al., 316 p.

NASCIMENTO MA, SILVA JRO (1998) Towards Historical R-trees. In: proceedings of the 1998 ACM Symposium on Applied Computing, pp 235-240

NEUGEBAUER H (1993)(ed.) Sonderforschungsbereich 350, Wechselwirkungen kontinentaler Stoffsysteme und ihre Modellierung, University of Bonn, Germany, 76p.

NEUMANN K (1987) Eine geowissenschaftliche Datenbanksprache mit benutzerdefinierbaren geometrischen Datentypen, PhD thesis, TU Braunschweig, Germany

NEUMANN K, LOHMANN F, EHRICH H-D (1992) An Experimental Geoscientific Database System. In: Vinken R (Ed.): Geologisches Jahrbuch, series A, vol. 122, Hannover, Germany, pp 91-100

NIEVERGELT J, HINTERBERGER H., SEVCZIK CK (1984) The GRID FILE: An Adaptable, Symmetric Multi-Key File Structure. In: ACM Transactions on Database Systems 9(1), pp 38-71

NIEVERGELT J, WIDMAYER P (1997) Spatial Data Structures: Concepts and Design Choices. In: Algorithmic Foundations of Geographic Information Systems, LNCS Tutorial No. 1340, Springer, Berlin et al., pp 153-197

NOACK R (1993) Verarbeitung dreidimensionaler geometrischer Objekte innerhalb des erweiterbaren Objekt-Management-Systems OMS, diploma thesis, Institute of Computer Science III, University of Bonn, 151p.

ODMG (1993) RGG Catell, The Object Database Standard: ODMG-93. Morgan Kaufmann Publishers, San Mateo, California, 169p.

OGIS I et al. (1997) Vision einer neuen Generation interoperabler GIS, techn. report, DFG joint project "Interoperable Geoscientific Information Systems"

OMG (1997a) CORBA 2.0/IIOP Specification. Formal Document 97-09-01, Object Management Group (http://www.omg.org/corba/csindx.htm)

OMG (1997b) CORBA Services: Common Object Services Specification. Formal Document 97-07-04, Object Management Group (http://www.omg.org/corba/csindx.htm)

OMG (1998) Object Management Group, the Common Object Request Broker: Architecture and Specification, Revision 2.2, Feb. 1998

ORENSTEIN JA (1986) Spatial Query Processing in an Object-oriented Database System. In: ACM SIGMOD Conference on Management of Data, ACM, München, Germany

PAPADIAS D, ARKOUMANIS D, KARACAPILIDIS N (1998) On the Retrieval of Similar Configurations. In: proceedings of the 8th Intern. Symposium on Spatial Data Handling, Vancouver, Canada, pp 510-521

PARENT C, SPACCAPIETRA S, ZIMANYI E, DONINI P, PLAZANET C, VANGENOT C (1998) Modelling Spatial Data in the MADS Conceptual Model. In: proceedings of the 8th Intern. Symposium on Spatial Data Handling, Vancouver, Canada, pp 138-150

PAUL H-B, SCHEK H-J, SCHOLL MH, WEIKUM G, DEPPISCH U (1987) Architecture and Implementation of the Darmstadt Database Kernel System. In: proceedings of ACM SIGMOD, San Francisco

PEUQUET DJ (1984) A Conceptual Framework and Comparison of Spatial Data Models. In: Cartographica, Vol. 21, No. 4, pp 66-113

PFLUG R, KLEIN H, RAMSHORN CH, GENTER M, STÄRK A (1992) 3D Visualization of Geologic Structures and Processes. In: LNES No. 41, Springer, Berlin et al., pp 29-39

PIGOT S (1992a) A Topological Model for a 3D Spatial Information System. In: proceedings of the 5th Intern. Symposium on Spatial Data Handling (SDH), Charleston, South Carolina, pp 344-360

PIGOT S (1992b) The Fundamentals of a Topological Model for a Four-Dimensional GIS. In: proceedings of the 5th Intern. Symposium on Spatial Data Handling (SDH), Charleston, South Carolina, pp 580-591

PILOUK M, TEMPFLI K, MOLENAAR M (1994) A Tetrahedron based 3D Vector Data Model for Geoinformation. In: proceedings of the Intern. GIS Workshop Advanced Geographic Data Modelling, AGDM'94, Delft

POLTHIER K, RUMPF M (1995) A Concept for Time-Dependent Processes. In: Visualization in Scientific Computing, Springer 1995, Berlin et al., pp 137-153

POSC (1997) www.posc.org

PREPARATA FP, SHAMOS MI (1985) Computational Geometry - An Introduction. Springer Verlag, Heidelberg

PREUSS H (1992) Mapping using Integrated Raster and Vector Data. In: Geolog. Jahrbuch, A122, Hannover, pp 167-176

PREUSS H, VINKEN H, VOSS HH (1991) Symbolschlüssel Geologie - Symbole für die Dokumentation und automatische Datenverarbeitung geologischer Feld- und Aufschlußdaten, 328 p.

QUERENBURG VB (1979) Mengentheoretische Topologie, 2nd edition. Springer, Berlin et al., 208p.

RAMSHORN C, KLEIN H, PFLUG R (1992) Dynamic Display for Better Understanding Shaded Views of Geologic Structures. In: Geolog. Jahrbuch A122, Hannover, pp 313--322

RAPER J (Ed) (1989) Three Dimensional Applications in Geographical Information Systems. Taylor & Francis, London 1989, 189p.

RAPER J, RHIND D (1990) UGIX(A): The Design of a Spatial Language Interface for a Topological Vector GIS, In: proceedings of the 4th Intern. Symposium on Spatial Data Handling, Zürich 1990, Vol. 1, pp 405-412

REQUICHA AAG, VOELCKER HB (1982) Solid Modelling: A Historical Summary and Contemporary Assessment, IEEE Comp. Graph. Appl., 2,2, Los Alamitos, CA, pp 9-24

REVERBEL F (1996) Persistence in Distributed Object Systems: ORB/ODBMS Integration. PhD Dissertation, Computer Science Department, University of New Mexico (http://www.ime.usp.br/reverbel/)

RHEINBRAUN (1991) Das rheinische Braunkohlenrevier, Rheinbraun AG Cologne, Germany, 6/91

RHIND DW, GREEN NPA (1988) Design of a Geographical Information System for a Heterogeneous Scientific Community. In: Intern. Journal of Geographical Information Systems, Vol.2, No.2

ROTH MT, SCHWARZ PM (1997) Don't Scrap It, Wrap It! A Wrapper Architecture for Legacy Sources. In: proceedings of the 23rd VLDB Conference, Athens, GR, http://www.ime.usp.br/reverbel/), pp 266-275

ROUSSOPOULOS N, LEIFKER D (1985) Direct Spatial Search on Pictorial Databases using Packed R-Trees. In: proceedings of ACM SIGMOD Intern. Conference on Management of Data

RUMBAUGH J, BLAHA M, PREMERLANI W, EDDY F, LORENSEN W (1991) Object--Oriented Modelling and Design. Prentice Hall, New Jersey, 391p.

SALGE F, SMITH N, AHONEN P (1992) Towards harmonized Geographical Data for Europe; MEGRIN and the Needs for Research. In: proceedings of the 5th Intern. Symposium on Spatial Data Handling, Charleston, SC, Vol. 1, pp 294-302

SAMET H (1990) The Design and Analysis of Spatial Data Structures; Addison-Wesley, Reading

SCHÄFER A (1994) Die Niederrheinische Bucht im Tertiär - Ablagerungs- und Lebensraum. In: Koenigswald W von, MEYER W (Eds.): Erdgeschichte im Rheinland. Fossilien und Gesteine aus 400 Millionen Jahren, Bonn, Germany, pp 155-164

SCHALLER J (1988) Das Geografische Informationssystem "ARC/INFO" und die mögliche Anwendung auf Geo-Daten, Fa. ESRI, Gesellschaft für Systemforschung und Umweltplanung mbH

SCHEK H-J, SCHOLL MH (1986) The Relational Model with Relation-Valued Attributes. In: Inform. Systems Vol. 11, No. 2, pp 137-147

SCHEK H-J, WATERFELD W (1986) A Database Kernel System for Geoscientific Applications. In: proceedings of the 2nd Symposium on Spatial Data Handling, Seattle

SCHEK H-J, WOLF A (1992) Cooperation between Autonomous Operation Services and Object Database Systems in a Heterogeneous Environment. In: Hsiao D, Neuhold EJ, Sacks-Davis R (Eds.), proceedings of IFIP TC2/WG2.6 Conference on Semantics of Interoperable Database Systems, DS-5, Lorne, Victoria, Australia

SCHEK H-J, WOLF A (1993) From Extensible Databases to Interoperability between Multiple Databases and GIS Applications. In: proceedings of SSD'93, Singapore, LNCS 692, Springer, pp 207-238

SCHNEIDER R, KRIEGEL HP (1991): The TR*-Tree: A new Representation of Polygonal Objects supporting Spatial Queries and Operations. In: Proceedings of the 7th Workshop on Computational Geometry. LNCS No. 553, Springer, 249-264

SCHOENENBORN I (1993) Verwaltung und Verarbeitung topologischer Strukturen in einem erweiterbaren Objektmanagementsystem. Diploma thesis, Institute of Computer Science III, University of Bonn, Germany

SELLIS T, ROUSSOPOULOS N, FALOUTSOS C (1987) The R+-Tree: A Dynamic Index for Multi-Dimensional Objects. In: proceedings of the 13th VLDB Conference, Grighton, pp 507-518

SHAMOS MI (1978) Computational Geometry. Ph. D. Thesis, Dept. of Computer Science, Yale University

SHEPHERD IDH (1991) Information Integration and GIS. In: Maguire et al. (1991), pp 337-360

SIEHL A (1988) Construction of Geological Maps based on Digital Spatial Models. Geol. Jb., A104, 2 Abb., Hannover, pp 253-261

SIEHL A (1993) Interaktive geometrische Modellierung geologischer Flächen und Körper. In: Die Geowissenschaften, vol. 11:10-11, 10 figs., Berlin, pp 343-346

SIEHL A, RÜBER O, VALDIVIA-MANCHEGO M, KLAFF J (1992) Geological Maps Derived from Interactive Spatial Modelling. In: From Geoscientific Map Series to Geo-Information Systems, Geologisches Jahrbuch, A (122), Hannover, pp 273-290

SMALLWORLD GIS (1996) Product Description, SMALLWORLD Systems GmbH

SNODGRASS R (1987) The Temporal Query Language TQUEL. In: ACM Transactions on Database Systems 12, pp 247-298

SNODGRASS R (1995) The TSQL2 Temporal Query Language. Kluwer Academic Publishers

SONNE B (1988) Raumbezogene Datenbanken für kartographische Anwendungen. In: Geo-Informationssysteme, 1(1), pp 25-29

SPACCAPIETRA S, PARENT C, ZIMANYI E (1998) Modelling Time from a Conceptual Perspective. In: proceedings of the Intern. Conference on Information and Knowledge Management, CIKM'98, Washington D.C., USA, Nov. 3-7

SPIES A, SPEVACEK R (1990) Implementierung von B*-Bäumen auf einem Datenbankkern für die Verwendung des Quadtree-Konzepts zur Approximation von Geometrien. Study thesis, Technical University of Darmstadt, Germany, 117p.

STONEBRAKER M, KEMNITZ (1991) The POSTGRES Next-Generation Database Management System. In: Communications of the ACM, 34:10, October

SVENSSON P, ZHEXUE H (1991) Geo-SAL: A Query LAnguage for Spatial Data Analysis. In: Günther O, Schek H-J (Eds.), proceedings SSD, LNCS No. 525, Springer, Berlin et al., pp 119-142

SYSTEM9 (1992) Systembeschreibung von SYSTEM9, COMPUTERVISION, BR9158, 4/92

SZYPERSKI C (1998) Component Software - Beyond Object-oriented Programming, Addison-Wesley, Essex, England, 411p.

THEODORIDIS Y, VAZIRGIANNIS M, SELLIS T (1996) Spatio-temporal Indexing for Large Multimedia Applications. In: proceedings of the 3rd IEEE Conf. on Multimedia Computing and Systems, pp 441-448

TOMLINSON RF (Ed.) (1972) Geographical Data Handling. IGU Commission on Geographical Data Sensing and Processing, Ottawa

TURNER AK (1992) Three-Dimensional Modelling with Geoscientific Information Systems, NATO ASI 354, Kluwer Academic Publishers, Dordrecht, 443p.

UML (1997) Harmon P, Watson M, Understanding UML: The Developer's Guide. Mogan Kaufman, San Francisco. http://www.rational.com

UTESCHER T, MOSBRUGGER V (1995) The Coexistence Approach - a Method for Quantitative Reconstructions of Tertiary Terrestrial Data using Plant Fossils, SFB series, No. 12, University of Bonn, Germany

VAN ECK JW, UFFER M (1989) A Presentation of System 9. In: Photogrammetry and Land Information Systems, Lausanne, March, pp139-178

VAN OOSTEROM P, VIJLBRIEF T (1991) Building a GIS on Top of the Open DBMS "Postgres". In: proceedings EGIS'91, pp 775-787

VAN OOSTEROM P, VERTEGAAL W, VAN HEKKEN M, VIJLBRIEF T (1994) Integrated 3D Modelling with a GIS. In: proceedings of the International GIS Workshop Advanced Geographic Data Modelling, AGDM'94, Delft

VIJLBRIEF T, VAN OOSTEROM P (1992) The Geo^{++} System: an Extensible GIS. In: proceedings of the 5th Intern. Symposium on Spatial Data Handling, 44-50, Charleston, SC, Vol. 1

VINKEN R (1988) Digital Geoscientific Maps -A Research Project of the Deutsche Forschungsgemeinschaft (German Research Foundation). In: Vinken R (Ed.), Construction and Display of Geoscientific Maps derived from Databases. Geol. Jahrbuch, proceedings of the Intern. Colloquium at Dinkelsbühl, FRG, Dec. 2-4, 1986, Geol. Jahrbuch, A104, Hannover, pp 7-20

VINKEN R (1992) From Digital Map Series in Geosciences to a Geo-Information System. In: Vinken R (Ed.), From Geoscientific Map Series to Geo-Information Systems. Proceedings of the Intern. Colloquium at Wuerzburg, Sept. 9-11, 1989, Geol. Jahrbuch, A122, Hannover, pp 7-26

VOGEL A, DUDDY K (1997) Java Programming with CORBA, Object Management Group. John Wiley & Sons, New York et al., 425 p.

VOIGTMANN A, BECKER L, HINRICHS KH (1996) Temporal Extensions for an Object-oriented Geodata Model. In: proceedings of the 7th Intern. Symposium on Spatial Data Handling, Delft

VOISARD A (1991) Towards a Toolbox for Geographic User Interfaces. In: proceedings SSD, LNCS No. 525, Springer, Berlin et al., pp 75-98

VOISARD A, SCHWEPPE H (1994) A Multilayer Approach to The Open GIS Problem. In: proceedings of the 2nd ACM workshop on Advances in Geographic Information Systems, Dec. 1-1, Gaithersburg, Maryland, USA

VOSS HH, OCHMANN M (1992) Fundamentals of Digital Geological Maps: Elements, Interrelationships, and Constraints. In: Geolog. Jahrbuch, A122, Hannover, Germany, pp 217-232

VOSS HH, MORGENSTERN D (1997) Interoperable Geowissenschaftliche Informationssysteme (IOGIS). In: GIS, 2/97, Wichmann-Verlag, Heidelberg, Germany, pp 5-8

WAGNER M (1998) Effiziente Unterstützung von 3D-Datenbankanfragen in einem Geo-ToolKit unter Verwendung vno R-Bäumen. Diploma thesis, Institute of Computer Science III, University of Bonn, 90p.

WALTHER J (1893) Einleitung in die Geologie als historische Wissenschaft. Beobachtungen über die Bildung der Gesteine und ihre organischen Einschlüsse. Vol. 1-3, Jena, 1055 p.

WATERFELD W (1991) Eine erweiterbare Speicher- und Zugriffskomponente für geowissenschaftliche Datenbanksysteme. Darmstadt PhD thesis No. D17, Technical University of Darmstadt, Germany, 188p.

WATERFELD W, BREUNIG M (1990) Kopplung eines Kartenkonstruktionssystems mit einem Geo-Datenbankkern. In: Pillmann W, Jaeschke A (Eds.), Informatik für den Umweltschutz, Informatik Fachberichte No. 256, Springer Verlag, Berlin et al., pp 344-354

WATERFELD W, BREUNIG M (1992) Experiences with the DASDBS Geokernel: Extensibility and Applications. In: From Geoscientific Map Series to Geo-Information Systems, Geolog. Jahrbuch, A (122), Hannover, Germany, pp 77-90

WATERFELD W, SCHEK H-J (1992) The DASDBS Geokernel - an Extensible Database System for GIS. In: Turner AK (Ed.), Three-Dimensional Modelling with Geoscientific Information Systems, proceedings of NATO Advanced Research Workshop, Santa Barbara, Kluver Academic Publisher, pp 69-84

WEDEKIND H (1974) On the selection of access paths in a database system. In: proceedings of IFIP Working Conference on Data Base Management 1974, North-Holland Publishing Comp., Amsterdam, pp 385-397

WEIKUM G, NEUMANN B, PAUL H-B (1987) Konzeption und Realisierung einer mengenorientierten Schnittstelle zum effizienten Zugriff auf komplexe Objekte. In: proceedings BTW, Darmstadt, Germany, pp 212-230

WEILER K (1988) The Radial Edge Structure: a Topological Representation for Non-manifold Geometric Boundary Modelling. Geometric Modelling for CAD Applications, Elsevier Science Publish., Oxford, pp 3-36

WIDMAYER P (1991) Datenstrukturen für Geodatenbanken. In: Vossen G, Witt K-U, (Eds), Entwicklungstendenzen bei Datenbank-Systemen, Oldenbourg, Germany, pp 317-362

WIDOM J (1995) Research Problems in Data Warehousing. In: proceedings of the 4th Intern. Conference on Information and Knowledge Management (CIKM), Nov.

WOLF A (1989) The DASDBS GEO-Kernel, Concepts, Experiences, and the Second Step. In: Buchman et al. (Eds.), proceedings of the 1st Symposium on Design and Implementation of Large Spatial Databases, Santa Barbara, CA, LNCS 409, Springer, Berlin et al., pp 67-88

WOLF A, DE LORENZI M, OHLER T, HAI NGUYEN V (1994) COSIMA: A Network Based Architecture for GIS. In: Nievergelt J, Roos Th, Schek H-J, Widmayer P (Eds.), IGIS'94: Geographic Information Systems, Monte Verita, Ascona, Schweiz, Feb. 28 - -March 4, LNCS 884, Springer, Berlin et al., pp 192-201

WORBOYS MF (1992) A Model for Spatio-Temporal Information. In: proceedings of the 5th Intern. Symposium on Spatial Data Handling, Charleston, SC, Vol. 1, pp 602-611

WORBOYS MF (1994) (Ed.), Innovations in GIS I. Taylor & Francis, London, 267p.

WORBOYS MF (1995) GIS - a computing perspective. Taylor & Francis, London, 376p.

WORBOYS MF, DEEN SM (1991) Semantic Heterogeneity in Distributed Geographic Databases. In: SIGMOD RECORD, Vol. 20, No. 4, December, pp 30-34

XU X, HAN J, LU W (1990) RT-Tree: An Improved R-Tree Index Structure for Spatiotemporal Databases. In: proceedings of the 4th Intern. Symposium on Spatial Data Handling, Zürich, Switzerland, Vol. 2, pp 1040-1049

ZHOU Q, GARNER BJ (1991) On the Integration of GIS and Remotely Sensed Data: Towards An Integrated System to Handle The Large Volume of Spatial Data. In: Proceedings of the 2nd Symposium SSD, LNCS No. 525, Springer, Berlin, pp 63-74

Index

Numerics

3D model data 152
3D object model 159
3D/4D
    geoinformation system 21, 103, 125
        component-based 103
α-shapes 65

A

access methods 24
algebraic topology 9
algorithm
    geometric 32
approximation
    in space and time 60
ARC/INFO 41

B

balanced restoration 121
building block 89

C

client-server connection 168, 169
commercial system 41
complexes
    simplicial 66
component
    connected 77
    for 3D/4D geoinformation system 125
component-based 6, 172
    geoinformation system 172
CORBA 169, 170, 173
coupling mechanism 168

D

DASDBS-Geokernel 5, 104
database design
    relational 30
    object-oriented 30
data
    integration 143
    modelling 57
    model integration 157
database
    access 92
    queries 88
        spatial 88
        temporal 88
    support 131

E

elementary geometric algorithms 32
equality of
    spatio-temporal objects 84
Erft block 137, 138

F

field-based
    approach 17
    model 13
    modelling 60

G

generalized map 63
geodata
    analysis 32
    management 22
    modelling 8

geodatabase 22
    kernel system 103
geoinformation science 3
geoinformation system 1, 41, 57
    component-based 103, 172
geological component 125
geologically defined geometries 137
geometric algorithm 32
geometry 77, 87
geo-objects
    in space 83
    in time 83
GeoStore 52, 53, 110
GeoToolKit 54, 109, 120, 161, 173, 176
GeO$_2$ 50
GEO^{++} 49
GIS
    raster-based 21
    research 3
    types 20
    vector-based 20
    3D/4D 21
GODOT 51
GRAPE 141, 142
grid file 28

H

hybrid GIS 21

I

index
    spatial 114
integrated application 155
integration
    data 143
    methods 143, 161
    systems 165
integrity checking component 126
integrity constraint
    spatial 86
    temporal 86
interactive geological modelling 131

L

layer model 19
Lower Rhine Basin 128, 132, 138, 143, 148

M

management of
    spatial and temporal objects 86
meta data approach 143
methods integration 143, 161
model
    field-based 13
    object-based 15
modelling
    field-based 60
    object-based 60
    of spatial and temporal objects 62
MR-Tree 95

O

object
    spatio-temporal 84, 98
object-based
    approach 17
    model 15
    modelling 60
Object Management System 107
object model 125
object-oriented database design 30
OMS 107
OMT 114
OODBMS 168, 171, 177
original data approach 144

P

pointset topology 9

Q

quadtree 24
queries 88

R

raster-based GIS 21
raster representation 17
relational database design 30
relationships
    spatial topological 69
    temporal topological 83

representations
    spatial 62, 153, 154
    temporal 62
research prototype 49
R-Tree 29, 95
R*-Tree 95

S

simplex 11
simplicial complex 10, 66, 67
    convex 66, 67
SMALLWORLD GIS 47
space 57
spatial
    access method 24, 115
    data model 17
    index 115
    representation 114, 153
    topological relationship 69
    topology 73
spatio-temporal
    attribute 89
    database access 92
    function 89, 90
        application 90
    join 91
    object 84, 98
    operation 89
    predicate 89
    selection 89
systems
    development 103
    integration 165, 166
SYSTEM 9 46

T

temporal
    change 73
    extension 17
    integrity constraint 86
    topological relationship 83
time 58
topological
    properties 58
    relationship 69, 83
topology 73, 77, 87

V

vector-based GIS 20
vector representation 17
visualization
    extensible 3D 119
    of spatial data 39
    of temporal data 39
    of spatio-temporal objects 98

W

well data 147, 152

Druck: Strauss Offsetdruck, Mörlenbach
Verarbeitung: Schäffer, Grünstadt

Lecture Notes in Earth Sciences

For information about Vols. 1–19
please contact your bookseller or Springer-Verlag

Vol. 20: P. Baccini (Ed.), The Landfill. IX, 439 pages. 1989.

Vol. 21: U. Förstner, Contaminated Sediments. V, 157 pages. 1989.

Vol. 22: I. I. Mueller, S. Zerbini (Eds.), The Interdisciplinary Role of Space Geodesy. XV, 300 pages. 1989.

Vol. 23: K. B. Föllmi, Evolution of the Mid-Cretaceous Triad. VII, 153 pages. 1989.

Vol. 24: B. Knipping, Basalt Intrusions in Evaporites. VI, 132 pages. 1989.

Vol. 25: F. Sansò, R. Rummel (Eds.), Theory of Satellite Geodesy and Gravity Field Theory. XII, 491 pages. 1989.

Vol. 26: R. D. Stoll, Sediment Acoustics. V, 155 pages. 1989.

Vol. 27: G.-P. Merkler, H. Militzer, H. Hötzl, H. Armbruster, J. Brauns (Eds.), Detection of Subsurface Flow Phenomena. IX, 514 pages. 1989.

Vol. 28: V. Mosbrugger, The Tree Habit in Land Plants. V, 161 pages. 1990.

Vol. 29: F. K. Brunner, C. Rizos (Eds.), Developments in Four-Dimensional Geodesy. X, 264 pages. 1990.

Vol. 30: E. G. Kauffman, O.H. Walliser (Eds.), Extinction Events in Earth History. VI, 432 pages. 1990.

Vol. 31: K.-R. Koch, Bayesian Inference with Geodetic Applications. IX, 198 pages. 1990.

Vol. 32: B. Lehmann, Metallogeny of Tin. VIII, 211 pages. 1990.

Vol. 33: B. Allard, H. Borén, A. Grimvall (Eds.), Humic Substances in the Aquatic and Terrestrial Environment. VIII, 514 pages. 1991.

Vol. 34: R. Stein, Accumulation of Organic Carbon in Marine Sediments. XIII, 217 pages. 1991.

Vol. 35: L. Håkanson, Ecometric and Dynamic Modelling. VI, 158 pages. 1991.

Vol. 36: D. Shangguan, Cellular Growth of Crystals. XV, 209 pages. 1991.

Vol. 37: A. Armanini, G. Di Silvio (Eds.), Fluvial Hydraulics of Mountain Regions. X, 468 pages. 1991.

Vol. 38: W. Smykatz-Kloss, S. St. J. Warne, Thermal Analysis in the Geosciences. XII, 379 pages. 1991.

Vol. 39: S.-E. Hjelt, Pragmatic Inversion of Geophysical Data. IX, 262 pages. 1992.

Vol. 40: S. W. Petters, Regional Geology of Africa. XXIII, 722 pages. 1991.

Vol. 41: R. Pflug, J. W. Harbaugh (Eds.), Computer Graphics in Geology. XVII, 298 pages. 1992.

Vol. 42: A. Cendrero, G. Lüttig, F. Chr. Wolff (Eds.), Planning the Use of the Earth's Surface. IX, 556 pages. 1992.

Vol. 43: N. Clauer, S. Chaudhuri (Eds.), Isotopic Signatures and Sedimentary Records. VIII, 529 pages. 1992.

Vol. 44: D. A. Edwards, Turbidity Currents: Dynamics, Deposits and Reversals. XIII, 175 pages. 1993.

Vol. 45: A. G. Herrmann, B. Knipping, Waste Disposal and Evaporites. XII, 193 pages. 1993.

Vol. 46: G. Galli, Temporal and Spatial Patterns in Carbonate Platforms. IX, 325 pages. 1993.

Vol. 47: R. L. Littke, Deposition, Diagenesis and Weathering of Organic Matter-Rich Sediments. IX, 216 pages. 1993.

Vol. 48: B. R. Roberts, Water Management in Desert Environments. XVII, 337 pages. 1993.

Vol. 49: J. F. W. Negendank, B. Zolitschka (Eds.), Paleolimnology of European Maar Lakes. IX, 513 pages. 1993.

Vol. 50: R. Rummel, F. Sansò (Eds.), Satellite Altimetry in Geodesy and Oceanography. XII, 479 pages. 1993.

Vol. 51: W. Ricken, Sedimentation as a Three-Component System. XII, 211 pages. 1993.

Vol. 52: P. Ergenzinger, K.-H. Schmidt (Eds.), Dynamics and Geomorphology of Mountain Rivers. VIII, 326 pages. 1994.

Vol. 53: F. Scherbaum, Basic Concepts in Digital Signal Processing for Seismologists. X, 158 pages. 1994.

Vol. 54: J. J. P. Zijlstra, The Sedimentology of Chalk. IX, 194 pages. 1995.

Vol. 55: J. A. Scales, Theory of Seismic Imaging. XV, 291 pages. 1995.

Vol. 56: D. Müller, D. I. Groves, Potassic Igneous Rocks and Associated Gold-Copper Mineralization. 2nd updated and enlarged Edition. XIII, 238 pages. 1997.

Vol. 57: E. Lallier-Vergès, N.-P. Tribovillard, P. Bertrand (Eds.), Organic Matter Accumulation. VIII, 187 pages. 1995.

Vol. 58: G. Sarwar, G. M. Friedman, Post-Devonian Sediment Cover over New York State. VIII, 113 pages. 1995.

Vol. 59: A. C. Kibblewhite, C. Y. Wu, Wave Interactions As a Seismo-acoustic Source. XIX, 313 pages. 1996.

Vol. 60: A. Kleusberg, P. J. G. Teunissen (Eds.), GPS for Geodesy. VII, 407 pages. 1996.

Vol. 61: M. Breunig, Integration of Spatial Information for Geo-Information Systems. XI, 171 pages. 1996.

Vol. 62: H. V. Lyatsky, Continental-Crust Structures on the Continental Margin of Western North America. XIX, 352 pages. 1996.

Vol. 63: B. H. Jacobsen, K. Mosegaard, P. Sibani (Eds.), Inverse Methods. XVI, 341 pages, 1996.

Vol. 64: A. Armanini, M. Michiue (Eds.), Recent Developments on Debris Flows. X, 226 pages. 1997.

Vol. 65: F. Sansò, R. Rummel (Eds.), Geodetic Boundary Value Problems in View of the One Centimeter Geoid. XIX, 592 pages. 1997.

Vol. 66: H. Wilhelm, W. Zürn, H.-G. Wenzel (Eds.), Tidal Phenomena. VII, 398 pages. 1997.

Vol. 67: S. L. Webb, Silicate Melts. VIII. 74 pages. 1997.

Vol. 68: P. Stille, G. Shields, Radiogenetic Isotope Geochemistry of Sedimentary and Aquatic Systems. XI, 217 pages. 1997.

Vol. 69: S. P. Singal (Ed.), Acoustic Remote Sensing Applications. XIII, 585 pages. 1997.

Vol. 70: R. H. Charlier, C. P. De Meyer, Coastal Erosion – Response and Management. XVI, 343 pages. 1998.

Vol. 71: T. M. Will, Phase Equilibria in Metamorphic Rocks. XIV, 315 pages. 1998.

Vol. 72: J. C. Wasserman, E. V. Silva-Filho, R. Villas-Boas (Eds.), Environmental Geochemistry in the Tropics. XIV, 305 pages. 1998.

Vol. 73: Z. Martinec, Boundary-Value Problems for Gravimetric Determination of a Precise Geoid. XII, 223 pages. 1998.

Vol. 74: M. Beniston, J. L. Innes (Eds.), The Impacts of Climate Variability on Forests. XIV, 329 pages. 1998.

Vol. 75: H. Westphal, Carbonate Platform Slopes – A Record of Changing Conditions. XI, 197 pages. 1998.

Vol. 76: J. Trappe, Phanerozoic Phosphorite Depositional Systems. XII, 316 pages. 1998.

Vol. 77: C. Goltz, Fractal and Chaotic Properties of Earthquakes. XIII, 178 pages. 1998.

Vol. 78: S. Hergarten, H. J. Neugebauer (Eds.), Process Modelling and Landform Evolution. X, 305 pages. 1999.

Vol. 79: G. H. Dutton, A Hierarchical Coordinate System for Geoprocessing and Cartography. XVIII, 231 pages. 1999.

Vol. 80: S. A. Shapiro, P. Hubral, Elastic Waves in Random Media. XIV, 191 pages. 1999.

Vol. 81: Y. Song, G. Müller, Sediment-Water Interactions in Anoxic Freshwater Sediments. VI, 111 pages. 1999.

Vol. 82: T. M. Løseth, Submarine Massflow Sedimentation. IX, 156 pages. 1999.

Vol. 83: K. K. Roy, S. K. Verma, K. Mallick (Eds.), Deep Electromagnetic Exploration. X, 652 pages. 1999.

Vol. 84: H. V. Lyatsky, G. M. Friedman, V. B. Lyatsky. Principles of Practical Tectonic Analysis of Cratonic Regions. XX, 369 pages. 1999.

Vol. 85: C. Clauser, Thermal Signatures of Heat Transfer Processes in the Earth's Crust. X, 111 pages. 1999.

Vol. 86: H. V. Lyatsky, V. B. Lyatsky, The Cordilleran Miogeosyncline in North America. XX, 384 pages. 1999.

Vol. 87: M. Tiefelsdorf, Modelling Spatial Processes. XVIII, 167 pages. 2000.

Vol. 88: S. Rodrigues-Filho, G. Müller, A Holocene Sedimentary Record from Lake Silvana, SE Brazil. XII, 96 pages. 1999.

Vol. 90: R. Klees, R. Haagmans (Eds.), Wavelets in the Geosciences. XVIII, 241 pages. 2000.

Vol. 91: I. Gilmour, C. Koeberl (Eds.), Impacts and the Early Earth. XVIII, 445 pages. 2000.

Vol. 92: P. C. Hansen, B. H. Jacobsen, K. Mosegaard (Eds.), Methods and Applications of Inversion. X, 304 pages. 2000.

Vol. 93: A. Montanari, Ch. Koeberl, Impact Stratigraphy. XIII, 364 pages. 2000.

Vol. 94: M. Breunig, On the Way to Component-Based 3D/4D Geoinformation Systems. XI, 199 pages. 2001.